# Teaching Engineering Made Easy

## A Friendly Introduction to Engineering Activities for Middle School Teachers

by Celeste Baine and Cathi Cox

Engineering Education
Service Center

Eugene, Oregon

# Teaching Engineering Made Easy

## A Friendly Introduction to Engineering Activities for Middle School Teachers

by Celeste Baine and Cathi Cox

Published by:
Bonamy Publishing
1004 5th Street
Springfield, OR 97477 U.S.A.
(541) 988-1005
www.engineeringedu.com



Printed in the United States of America

Teaching Engineering Made Easy: A Friendly Introduction to Engineering Activities for Middle School Teachers

ISBN 10: 0-9711613-0-5
ISBN 13: 978-0-9711613-0-6

**How to Order:**
Single copies may be ordered from:
Engineering Education Service Center,
1004 5th St., Springfield, OR 97477;
telephone (541) 988-1005

CIP Pending

# Contents

# MATH AND SCIENCE GIVE YOU MORE OPTIONS

# What you can do with this book...

This easy and exciting, time and work saving book was developed to help middle school teachers with no engineering background teach engineering. The activities are designed to help teachers stimulate student's thought processes and get them thinking like an engineer. By using this teaching guide, teachers can help students see that engineering is not something to be afraid of but a realistic and practical way to solve the problems of everyday life.

*Teaching Engineering Made Easy* gives classroom teachers an easy and dynamic way to meet curriculum standards and competencies. The lessons and activities actively engage students in learning about engineering and our technological world by applying creativity and innovation as they complete the projects. The activities were developed and tested by a seasoned science teacher who sought to expose students to engineering with enjoyable learning experiences.
(For easy use, the lessons and activities are spiral bound to enable an 8 1⁄2" x 11" lay-flat format for photocopying.)

## This book is organized into 3 sections:

**1. Introduction to Engineering**
**2. Engineering Activities**
  **a. Team building activities**
  **b. Problem solving activities**
  **c. Chemical, mechanical and civil engineering activities**
  **d. Engineering design activities**
**3. Engineering Puzzles**

**Each section offers detailed lessons with reproducible student activity sheets. Each lesson can include:**

- **Teacher notes**
- **Background information**
- **Standards alignment**
- **A list of materials needed to complete the activity**
- **An easy-to-follow procedure for presenting the lesson**
- **Student sheets**
- **Safety notes**

# Quick View - National Science Education Standards
# Linking Specific Activities to the Standards

| | Content Standard A Science as Inquiry | | Content Standard B Physical Science | | Content Standard E Science and Technology | | Content Standard G History and Nature of Science | | |
|---|---|---|---|---|---|---|---|---|---|
| | Abilities Necessary to do Scientific Inquiry | Understanding Scientific Inquiry | Properties and Changes of Properties in Matter | Motions and Forces | Abilities of Technological Design | Understanding Science and Technology | History of Science | Science as a Human Endeavor | Nature of Science |
| **Whale Band-Aid** | X | X | | | | | | X | |
| **Carbon Copy Creations** | X | | | | X | X | | X | X |
| **Digging nto Diapers** | X | | X | | | | | X | X |
| **The Bountiful Bag** | X | X | X | | | | | X | X |
| **The Pop in your Pop** | X | X | X | | | | X | X | X |
| **Reactionary Rockets!** | X | X | X | X | | X | | X | X |
| **Stretching the Truth** | X | X | X | | | | | X | X |
| **The Mark of Success** | X | X | | X | X | X | | X | X |
| **How Do I Hover?** | X | X | | X | X | X | X | X | X |
| **Totally Tops** | X | X | | X | X | X | | X | X |
| **The Indy Card Car** | X | X | | X | X | X | | X | X |
| **The Cardboard Chair** | | | | | X | X | | X | X |
| **Bolstering Books** | | | | | X | X | | X | X |
| **Pasta Bridges** | | | | | X | X | | X | X |
| **Ten Second Tower** | | | | | X | X | | X | X |
| **Two Feet Feat** | | | | | X | X | | X | X |
| **Boat Bonanza** | X | X | | X | X | X | X | X | X |

| Learning Experience | | Skills and Main Concepts Developed in Each Learning Experience |
|---|---|---|
| 1. | The Whale Band-Aid | Problem solving, critical thinking, communication. |
| 2. | Carbon Copy Creations | Critical thinking, communication, recognizing patterns |
| 3. | Creating A Collaborative Clan | Communication, critical thinking |
| 4. | What's in the Bag? | Critical thinking, communication |
| 5. | Hampered By Height | Problem solving, critical thinking, measurement, communication, procedure design |
| 6. | The Puzzle of Nine | Critical thinking, problem solving, mass measurement, procedure design |
| 7. | Digging Into Diapers | Predicting, inferring, observing, investigating, physical/chemical properties, polymers |
| 8. | The Bountiful Bag | Observing, predicting, hypothesizing, experimental design, investigating, physical/chemical properties, physical/chemical changes, chemical reactions, acid/bases/pH (as appropriate), chemical formulas and equations (as appropriate) |
| 9. | The Pop In Your Pop | Observing, investigating, acids/bases/pH, solutions, chemical reactions, chemical formulas and equations |
| 10. | Reactionary Rockets | Observing, inferring, measurement, experimental design, investigating, chemical reactions, reaction rates, motion |
| 11. | Stretching the Truth | Measurement, collecting and organizing data, graphing, graphical analysis, experimental design, investigating, variables, polymers, elasticity |
| 12. | The Mark of Success | Observing, procedure design, data collection, measurement, accuracy, precision |
| 13. | How Do I Hover? | Observing, critical thinking, problem solving, investigating, forces, motion, friction, product design |
| 14. | Totally Tops | Observing, measuring, collecting data, critical thinking, problem solving, investigating, forces, product design |
| 15. | The Indy Card Car | Critical thinking, problem solving, experimenting, measuring, collecting data, graphing, graphical analysis, simple machines, speed/ momentum, force, motion, friction, product design |
| 16. | The Cardboard Chair | Critical thinking, problem solving, forces, center of mass, stability, product design |
| 17. | Bolstering Books | Critical thinking, problem solving, forces, stability, beam strength, column strength, product design |
| 18. | Pasta Bridges | Critical thinking, problem solving, observing, torque, center of mass, measurement, compression, tension, beam strength, product design |
| 19. | Ten Second Tower | Critical thinking, problem solving, observing, measurement, center of mass, tension, compression, forces, beam strength, column strength, stability, product design |
| 20. | The Two Foot Feat | Critical thinking, problem solving, observing, measurement, center of mass, tension, compression, forces, beam strength, column strength, stability, product design |
| 21. | Boat Bonanza! | Observing, inferring, predicting, critical thinking, problem solving, mass, measurement, investigating, experimenting, collecting data, graphing, graphical analysis, density, buoyancy, forces, product design |

# The New Face of Engineering

You're a progressive thinker, right? You've let go of the stereotypes about women not doing traditional "male" jobs and math geniuses breaking the fashion rules with pocket protectors. So why are you still clinging to the idea that engineering is all about building bridges and skyscrapers?

Today's engineering majors might go on to find themselves in any of the following scenarios:

- A test engineer crashing expensive sports cars into walls to make them safer
- A forensic engineer evaluating crime scene evidence to narrow down the search for a criminal
- A design engineer creating a robot that can save people from burning buildings
- A pharmaceutical engineer discovering a cure for a disease that has killed millions
- A financial engineer analyzing Wall Street's patterns to try to predict future trends

Engineers strive to make our lives better, easier, cheaper, more efficient and more fun by solving problems in everyday life. Engineers are practical inventors. It is through the work of engineers that we are able to have camera phones, wireless computers, satellite TV, airplanes, hydrogen powered cars, digital music, underwater robots, air conditioning, indoor plumbing, cosmetics, titanium knee and hip replacements, and the list goes on and on. Almost everything you touch has been influenced or designed by an engineer directly or indirectly. It is impossible to think of a major technical development that hasn't included the work of engineers. Many internationally famous companies such as Hewlett Packard, Intel and Apple wouldn't exist if one or more practical inventors (engineers!) didn't get together and make it happen. With solid roots in engineering, these companies have grown in epic proportions.

**Engineering is a way to make life better.** Many problems are solved by applying math principles, but math is just one tool in the engineer's toolbox. Inspiration, creativity, imagination, energy, passion and communication skills are also extremely important.

## Engineers Can Do Anything!

Engineers are modern day superheroes. They design, invent, build and concoct the most remarkable physical achievements known to humanity, and they are some of the most creative people on earth. If you have ever wanted to reduce pollution, end world hunger, become president of the United States (three presidents were engineers), improve the environment, invent exciting technology, become an astronaut, design race cars, solve complex problems, or be on the cutting edge in a dynamic career, then engineering has something for you.

Not every engineer craves the traditional engineering path. Diverse and plentiful opportunities exist for the educated non-mainstream engineer. It's a well-kept secret, but many engineering graduates don't go straight into engineering. Many go on to law school to become patent or environmental attorneys. Many go on to medical school, many find their way to Wall Street to become financial engineers and still more become technical writers or teachers. A 1995 NSF survey found that only 38 percent of those in the U.S. work force with a degree in engineering actually work as engineers. Another four percent say they work in a related science field, and an additional 48 percent aren't considered engineers or scientists, but say their work is related to engineering. There are many opportunities for the engineer in search of alternatives to the standard career, because engineers can do anything.

## Engineers are Creative?

Most people don't label engineers as creative, but employers such as Disney, Pixar, Nike, Intel, Hasbro, Mattel, Industrial Light and Magic (Lucas Films), Nintendo, Sony, Hewlett Packard and the millions other companies that employ engineers might argue with you. Engineers are the concept people and often the idea people too. They are the ones that decided they didn't like to clean house so now we have the Roomba vacuum cleaner that automatically vacuums your house. They are the ones that figured out how to make a roller coaster careen forward at 120 M.P.H. in four seconds without killing you. They are the ones that figured out how to make cars that can run on electricity or fuel cell technology to keep our atmosphere cleaner. They create medical equipment used by doctors to keep us healthy and they even work in the food industry to make chocolate taste better.

***Engineers put the pieces together and solve the problems of our day.***

The American Society of Civil Engineers obviously also sees engineers as creative because they sponsor a "Concrete Canoe" competition for undergraduate students. The canoe must be made entirely out of concrete and in addition to that, it has to float and even win a race against other concrete canoes created by students at other institutions. Creativity is certainly an essential element in this arena.

Some of the most amazing inventions and technologies on the market today exist because one engineer had an idea. Look back at old pictures of the bicycle. People wanted it to go faster, they wanted it to go down mountains and wanted it to be more comfortable. The difference is engineering. Each year, engineers went back to the drawing board and the bicycle got better. What will bicycles look like in another ten years? It's up to today's students and their imagination to tell us. Today's students will make the world a better place, a place that is safer, more fun and enjoyable for everyone.

A practical knowledge of business is becoming increasingly important as more engineers become involved with start-up companies. Engineering schools help students develop entrepreneurial skills such as dealing with venture capitalists, writing business plans, and understanding the business processes of a company. Students have to understand not only what it takes to technically create an invention but the financial and marketing aspects of their invention as well.

Engineers of all types have a **wildly** successful history of entrepreneurial ventures. Some engineers seem to be born inventors. They seem to have innate knowledge about how to be more efficient, make things easier to use, create new products, or save money. Someone (probably an engineer) invented most of the products or items that you can see, touch, smell and hear.

## Merging Science and Art

*"By working hard at developing my skills as an engineer, and being willing to say five simple words, the whole world becomes your oyster. The words? "Yeah, I can do that". Those five words and the ability to back it up is what gets you challenging work, fine homes and fun cars to drive."*

– Kyle Milliken, Project Engineer

Kyle Milliken, a mechanical engineer for an international manufacturing facility in Huizhoua China said, "To me, engineering is not a job but, a way of seeing things and can lead to literally a world of opportunities. This way of seeing things or perspective is the key ingredient to being an engineer. Walt Disney wants a new thriller ride to delightfully terrify visitors at the park, some person grabs a piece of paper and starts sketching. A million pounds of aluminum, jet fuel, people and tasteless food goes rumbling down the runway and takes flight, it's because some person had an idea. A doctor inserts a tiny device into the beating heart of a patient and his or her life becomes better, it's because somebody had a concept. Engineers are those people. Engineers turn ideas into reality. The vision, the ideas and the concepts are the important part, math is just a tool that engineers use to do it."

Engineering is a rigorous major. There is no doubt that the pursuit of an engineering degree will challenge, frustrate and confound you. As in anything, to become one of the best takes perseverance, self discipline, and enormous amounts of effort, concentration and a keen focus on your goal. You have to know that your goal will be worth it in the end, and that you are doing something that gives you a sense of satisfaction and happiness.

Theme parks are venues that attempt to merge science and art to create illusion that is so close to reality your mind and emotions cannot distinguish between the real and the magical. The idea with a theme park is to envelop the visitor in a seemingly different time and place based on stories.

A good story weaves a complex web, completely immersing the viewer through sight, hearing, smell, taste, and touch. The more senses that are enveloped simultaneously, the more real the created environment appears to be. To have a make-believe environment seem real, the technology behind it must be invisible. Art and science must blend into illusion. The Imagineers, (as Disney calls them) are a team of people that are responsible for the creation and development of all elements of a theme park. They know they have done their job well when their guests return again and again to see the magic they have created.

According to Nathan Naversen, a former employee of ITEC Entertainment, "Engineers figure out a way to make it work, whether it be sizing the structural columns and measuring shear forces on a roller coaster, or developing new electronics to make an animatronic character function. Engineers do the math to make everything 'stand up.'"

Other industries that hire engineers to design, build, conceive, produce or orchestrate entertaining environments include themed retail stores, themed restaurants, sports facilities, museums, aquariums, zoos, casinos, cinemas, expos, recreational facilities, coliseums, and planetariums.

# Why Teach Engineering?

Children, just like engineers, are creative, innovative, and imaginative when it comes to solving problems. In engineering design, there are usually multiple ways to solve a problem. Different solutions fit different societal needs at different times. For example, in the 1970's gasoline was inexpensive so the cars were heavier and had larger engines. People enjoyed the get-up-and-go. Today gasoline is more expensive so engineers work to design cars that are lighter and more fuel efficient. Some cars today even run on electricity or alternative fuels such as hydrogen because of the work of engineers.

**Engineering is problem solving.** Many teachers enjoy teaching it because it combines math and science lessons, team building and creativity with a practical twist. Students learn to work together, increase their communication skills and enhance their presentation abilities by demonstrating and discussing their design strategy with the rest of the class.

Hands-on activities or project-based learning are fun and effective ways to help students retain more math and science concepts. By choosing to teach engineering, teachers can help students make the links between the classroom and their everyday lives. Project based learning can help students visualize abstract science and math concepts. Engineering design serves as the bridge to bring color to math and science concepts that make our designed world more understandable.

# Tremendous Job Growth in Engineering

According to the NSF publication *Science and Engineering Workforce*, the projected demand for Science and Engineering (S&E) workers from 2000-2010 is expected to increase about three times faster than the rate for all occupations. Although the economy as a whole is expected to provide approximately 15 percent more jobs over this decade, employment opportunities for S&E jobs are expected to increase by about 47 percent (about 2.2 million jobs).

Approximately 86 percent of the increase in S&E jobs will likely occur in computer–related occupations. Overall employment in these occupations across all industries is expected to increase by about 82 percent over the 2000-2010 decade, adding almost 1.9 million new jobs. The number of jobs for computer software engineers is expected to increase from 697,000 to 1.4 million, and employment for computer systems analysts is expected to grow from 431,000 to 689,000 jobs.

Within engineering, environmental engineering is projected to have the biggest relative employment gains, increasing by 14,000 jobs, or about 27 percent Computer hardware engineering is also expected to experience above-average employment gains, growing by 25 percent. According to the department of labor, the number of biomedical or medical-related engineering jobs will increase by more than 30 percent in the next five years. However, students entering the science and engineering workforce today made decisions 14 years ago that enabled them to follow a course of study that would make them qualified applicants. Students making those same decisions in middle school today will not complete their advanced training for science and engineering jobs until 2018 or 2020. It is more important than ever that students have a good understanding of what engineers do so that they can make informed choices.

*Fundamentally, interest in engineering has to start at elementary or middle school to get students on the right track so they can take the right classes in high school. According to the 2005 National Data Profile by Achieve, Inc. Preparation for postsecondary education and good jobs begins well before high school. Students who take challenging courses and meet high standards in middle school are much more likely to enter high school ready to succeed. Algebra is widely recognized as a "gateway" course — students who take it by the end of 8th grade are much more likely to take rigorous courses in high school that lead to a college degree.*

*The findings are similar in high school aged students. In this case,AdvancedPlacement (AP) classes serve as the gateway courses. Research has shown that a powerful predictor of whether high school students will graduate and earn a college degree isn't only grades or even test scores, but rather the rigor of the high school curriculum they complete. Taking a high-level math course beyond Algebra II (trigonometry or calculus) is a key indicator of such a curriculum. Juniors and seniors taking AP exams or gateway courses in 2003 was only 11.4 percent.*

**With little formal K-12 education in engineering, students may not be prepared to chart their course into engineering school.**

The National Science Board, the independent board that advises the president and Congress on matters of science and engineering issued a report that the decline in S&E is due to a lack of student interest.

> *We have observed a troubling decline in the number of U.S. citizens who are training to become scientists and engineers, whereas the number of jobs requiring science and engineering training continues to grow,* the board wrote in an introduction to its Science and Engineering Indicators 2004.

A separate report from the U.S. Education Department's National Center for Education Statistics said that more than half of U.S. students are not taking any science in their senior year of high school. The Education Department's report was based on transcripts of more than 20,000 graduating high school seniors from 277 public and private high schools.

The National Science board also points out that *if action is not taken now to change these trends,* the board said, *we could reach 2020 and find that the ability of U.S. research and education institutions to regenerate has been damaged and that their preeminence has been lost to other areas of the world.*

## An Achievable Goal

A fundamental problem with students being interested in engineering is that many don't understand what engineering is, what engineers do, how it impacts lives and how it can be a rewarding career. An excellent goal is to ensure that every student entering high school is exposed to engineering and has the resources (mentors, books, videos, etc.) available to make an informed choice on whether to pursue engineering or not.

## Easy Ways to Talk to Kids About Engineering

1. Make Hollywood associations. Kids today are over stimulated. From an early age they could watch cartoons 24/7. The special effects in movies and on television are far more exciting or stimulating than just 10 years ago. Information via the Internet is brought to them constantly and immediately. They are bombarded by the media with images and solutions to every problem.

The image of the engineer portrayed by the media isn't the best. We see doctors and lawyers on TV all the time but where are the engineers? To get the kids of today interested in a career, we have to make the pop culture links.

**As an assignment to get the students warmed up, have them watch a movie that you select and identify the engineers doing engineering work. Two examples are listed below. See how many more you can come up with.**

a. Show students how Dr. Octopus in Spiderman was an engineer instead of a mad scientist. Have students build a list of all the reasons that Dr. Octopus was an engineer (Peter Parker is also an engineer but they never call him one).

b. Willy Wonka was a chemical engineer. Have students watch this movie in class or as a homework assignment and identify scenes in which he did engineering work.

2. Find out what students like and link it back to engineering. This is a great way to connect with students because you can link anything to engineering!
    a. Do they like music? If so, have them research and write a paper about the engineers that developed the iPod. Or, have them give a presentation about how CDs are made.
    b. Do they like skateboarding or sports? Bring a skateboard to class and have the students think about the design of the trucks, bearings or board. Show what things a skateboard engineer would do. How would your students do it differently? More information on engineers and sports equipment can be found in High Tech Hot Shots: Careers in Sports Engineering by Celeste Baine ($19.95).
    c. Do they like to eat? Show them how engineers work at Hershey, Mars, and other chocolate factories developing delicious chocolate treats.
    d. Do they enjoy video games?
    e. Do they enjoy watching movies?
    f. Do they enjoy using a cell phone?
    g. Do they like to chat online?

Much of the information for students to give reports on their favorite products is available at How Stuff Works (www.howstuffworks.com). If a student is particularly interested in video games, it is always good to have students include an employer's (Nintendo, Playstation, EA games, etc.) job posting for an engineer. You will find that many postings require excellent communication skills as well as technical ability. For example, to become an iPod engineer requires technical ability, creativity, communication skills, passion, energy and the ability to work in a team (www.apple.com). Most students don't realize that being a well-rounded person is a good skill for engineers!

## How Do Students Prepare for Engineering while in High School?

Obviously, academic preparation is essential to exploring engineering as a career. In high school, classes in algebra, trigonometry, biology, physics, calculus, chemistry, computer programming, or computer applications can be excellent indicators of aptitude and determination to study engineering. While all of these courses are not required to get into every engineering school, early preparation can mean the difference between spending four or six years in college. Some universities also require two to three classes in a foreign language for admission. Advance placement or honors courses always help, as do high ACT and SAT scores.

But grades aren't the only thing that matter. Students should get involved in engineering-related extracurricular activities to not only beef up their college application, but also to decide if they will enjoy the field. Volunteering, interning or acquiring a summer job working in a lab, a pharmacy, a manufacturing facility or an architectural/engineering firm is very valuable. Encourage students to not be afraid to try a few different things until they find one they like. The possibilities are endless!

## What are the Current Trends in Engineering?

With new advances in technology developing daily, every facet of engineering is evolving on a regular basis. Perhaps this is most prevalent in biomedical engineering, a subcategory in which industry professionals design instruments, software, and medical devices to help physicians, nurses, veterinarians, therapists and technicians solve complex medical problems and enhance the quality of medical care.

***With new advances in technology developing daily, every facet of engineering is evolving on a regular basis.***

Imagine designing a medical device that restores functions to someone who has lost them to disease or injury. The pacemaker was invented by biomedical engineers who literally gave recipients the ability to perform physical activities, such as climbing a flight of stairs or walking around the block that they were previously unable to do. A few years ago, four engineering students at Northwestern built a prosthetic device that allowed a burn victim who had lost a hand in a fire to play tennis again.

Another hot trend in engineering is mechatronics (mechanics and electronics). In technical terms, it's the *"synergistic integration of mechanical, electrical and computer systems."* In non-techie terms: You get paid to play with toys!

According to Helene Brooks, the Director of Public Affairs for Vaughn College of Aeronautics and Technology, the college will be offering its first Bachelor of Science engineering program in mechatronics in 2007.

*"The goal of this program is to provide students with the fundamental knowledge of mechanical and electronic engineering and enable them to design smart engineering components that exhibit precise performance,"* she says. *"Graduates of this program will acquire knowledge in the areas of mechanical engineering, electronics, computers, controls theory and design processes to create smart and more functional and adaptable products."*

Because most modern gadgets are embedded with electronic control systems, mechatronics engineers are involved in the design and construction of almost everything electronic: computers, appliances, robots, anti-lock brakes, you name it! The good news is, there are endless possibilities for a career in the field, and unless the world suddenly regresses to a pre-Industrial Revolution state, they're just going to increase. The bad news is, these engineers might have so much fun at their job that they'll become workaholics!

Whatever students are interested in, medicine, robotics, architecture, design or science, there's probably an engineering job with their name on it. Of course, by the time they graduate from college, there could be a whole new subcategory of engineering. They could be working on armies of robots, skyscrapers on Mars, or even brain transplants.

## Girls in Engineering

Girls make great engineers! Just ask these influential women about their contributions to the field and they will tell you that there is nothing a girl cannot do!

***Ada Byron Lovelace*** - collaborated with Charles Babbage, the Englishman credited with inventing the forerunner of the modern computer - wrote a scientific paper in 1843 that anticipated the development of computer software (including the term software), artificial intelligence, and computer music. The U.S. Department of Defense computer language, ADA, is named for her.

***Amanda Theodosia Jones*** - invented the vacuum method of food canning, a process that completely changed the entire food processing industry. In a move typical of women inventors of the 19th century, Jones denied the idea came from her inventiveness, but rather from instructions received from her late brother from beyond the grave.

***Ellen Swallow Richards*** - pioneered the field of environmental engineering with her groundbreaking research into water contamination. In 1870, she helped conduct the first analysis of Massachusetts' water supply and led the research on two subsequent testings. The work set the standard for the United States and the world. She showed incredible foresight with her insistence that the earth's environment be examined as a whole, rather than in bits and pieces. She also urged tighter controls over solid waste disposal and air, food, and water purity.

***Mary Engle Pennington*** - revolutionized food delivery with her invention of an insulated train car cooled with ice beds, allowing for the first time the long-distance transportation of perishable food.

***Mary Anderson*** - invented the windshield wiper in 1903. By 1916, they were standard equipment on all American cars.

***Beulah Louise Henry*** - known as "the Lady Edison" for the many inventions she patented in the 1920's and 1930's. Her inventions included a bobbinless lockstitch sewing machine, a doll with bendable arms, a vacuum ice cream freezer, a doll with a radio inside, and a typewriter that made multiple copies without carbon paper. Henry founded manufacturing companies to produce her creations and made an enormous fortune in the process.

***Hedy Lamarr*** - the 1940's actress known for her line "Any girl can be glamorous. All you have to do is stand still and look stupid" invented a sophisticated and unique anti-jamming device for use against Nazi radar. While the U.S. War Department rejected her design, years after her patent had expired, Sylvania adapted the design for a device that today speeds satellite communications around the world. Lamarr received neither money, recognition, nor credit.

***Grace Murray Hopper*** - a Rear Admiral in the U.S. Navy, developed the first computer compiler in 1952 and originated the concept that computer programs could be written in English. She once remarked, "No one thought of that earlier because they weren't as lazy as I was." Hopper is also the person who, upon discovering a moth that had jammed the works of an early computer, popularized the term "bug." In 1991, Hopper became the first woman, as an individual, to receive the National Medal of Technology. One of the Navy's destroyers, the U.S.S. Hopper, is named for her.

***Stephanie Kwoleks*** - discovered a polymade solvent in 1966 that led to the production of "Kevlar," the crucial component used in canoe hulls, auto bodies, and perhaps most importantly, bulletproof vests.

***Ruth Handler*** - best known as the inventor of the Barbie doll, also created the first prosthesis for mastectomy patients.

***Dr. Bonnie J. Dunbar*** - Helped to develop the ceramic tiles that enable the space shuttle to survive re-entry. In 1985, she had an opportunity to test those tiles first hand, as an astronaut aboard the shuttle.

***Elsa Garmire*** - made tremendous advances in optical devices and quantum electronics that made the commercial use of lasers feasible. Garmire discovered and explained key features of light scattering and self-focusing, and a host of other phenomena crucial to optical technology.

## Female Representation

- Only 19 percent of the science, engineering, and technology workforce is female.
- By the eighth grade, twice as many boys as girls show an interest in science, engineering, and mathematics careers.
- Fewer girls than boys enroll in computer science classes, feel self-confident with computers, or use computers outside the classroom.
- The percentage of women graduating with computer science degrees has decreased 25 percent.
- By the eighth grade, girls' interest in mathematics and confidence in their mathematics abilities have eroded, even though they perform as well as boys in this subject.
- Nearly 75 percent of tomorrow's jobs will require use of computers; fewer than 33 percent of participants in computer courses and related activities are girls.

Brianna Bethel of University of Colorado suggests that *"In order to be effective in promoting engineering to girls, you must first understand the main reasons why more women are not studying engineering and science.* She goes on to give the following reasons:

1. Many studies have shown the largest factor may be an incorrect perception of engineers. For children growing up with an engineer as a parent, relative, or family friend, engineers are viewed as people helping solve problems related to everyday life. This view is not shared by the majority of society. According to a Harris Poll on public perceptions of engineering, almost 60 percent of women in the United States do not know what engineers do for our society.

2. Another huge problem daunting women in engineering is the negative stereotypes surrounding engineers combined with a lack of positive role models. Starting at a young age, many children are introduced to science and engineering through television and cartoon shows. For example, Dexter's Laboratory encourages the image that men are superior intellectually. Dexter, the main scientist, is often interrupted by his not-so-bright sister. Recent shows like Bill Nye: the Science Guy and Beakman's World offer a way to inform children of science, but almost all of these shows feature weird looking men as the main scientists. If women are portrayed in the show, they are often lab assistants or students learning from a male scientist. One new show on PBS called DragonFly TV has both a male and female host. Each episode features different children performing different science activities. This is a start in the right direction.

3. Another problem with the media's portrayal of engineers is the negative images associated with engineering. Oftentimes, the media projects scientists as wearing huge glasses, looking dorky, and having no social skills. These negative images and stereotypes continue as girls get older. Even magazines such as Time use demeaning titles such as "Nerd Within" and "Could There Be a Geek Strike in Gates' Future?" The negative images inundate primetime television as well. In survey of fifteen hundred television viewers, researchers found that the more people watch television, the more they think scientists are odd and peculiar.

Media such as this has negative effects on both boys and girls. However, for girls, the negative result is more detrimental due to the lack of female role models in engineering. Upon analyzing television shows from 1994 to 1999, researchers from the Department of Commerce found that 75 percent of the scientists on primetime television were white males. This accurately reflects society, in 2000, only 10.6 percent of the engineering workforce was female. With proportions such as these, girls often cannot imagine themselves fitting into a 'man's occupation'. - Reprinted with permission from Brianna Bethel

## WHY AREN'T THERE MORE WOMEN INVENTORS IN THE RECORD BOOKS?

In the early years of the US, a woman could not get a patent in her own name. A patent was considered property, and until the 1800's, women could not own property in most states. Instead, the patent had to be registered in the woman's father's, brother's, or husband's name.

## What Can We Do?

According to the Advocates for Women in Science, Engineering, and Mathematics (AWSEMS), *Evidence overwhelmingly suggests that girls beginning their exploration of science and math as well as women who have already achieved high career goals in these fields benefit tremendously from vertical, dynamic mentoring networks, that is,* ***from mentoring relationships*** *involving a more experienced individual and a less experienced individual in which both profit from the insights, experiences, and enthusiasm of the other.*

There are many ways to find mentors. Encourage students to consider questions such as: Who can they talk to about their career? Who will take a special interest in their goals? Who do they admire? Who do they want to model themselves after or who do they want to emulate? Have them try to pick a mentor in their field of interest, but don't be limited by this approach. The most important thing is that they find someone they respect, admire, and can talk to easily.

Another approach to having your students find out if they will like engineering is to join a science or engineering club. This can be a way to have fun with math and science, meet other girls interested in those subjects, and learn about amazing careers. A club can provide girls with hands-on science, engineering, and math activities that are both fun and educational. Club members can also visit sites to meet and learn about women in science, math, and engineering careers.

If your school doesn't have a club, consider starting one. Look for instructions at the AWSEM Web site at www.awsem.com.

## Biomedical Engineering the Top Choice for Women

Biomedical engineering is, in a very real sense, people engineering. The objective of biomedical engineering is to enhance health care by solving complex medical problems using engineering principles. Those who specialize in this field want to serve the public, work with health care professionals, and interact with living systems. It is a broad field that allows a large choice of sub-specialties. **Many students say they choose biomedical engineering because it is people-oriented.**

Women seem naturally drawn to biomedical engineering. Maybe it is its social utility or maybe it satisfies the caregiver instinct. Whatever the case, according to the American Society for Engineering Education, biomedical engineering leads all engineering disciplines in the percentage of degrees awarded to women at all levels — bachelor's, master's and doctoral. Forty-five percent of all biomedical engineering degrees at the bachelor level in 2004 went to women.

Why are women attracted to biomedical engineering? For Pennsylvania State University graduate student Janice Turlington, the most attractive aspect of biomedical engineering is *"...the ability, or at least the opportunity, to be very creative. In other avenues of engineering, there's more of a rigidity to it. Biomedical engineering is really brand new. There are no cut and dry answers."*

*"I think what we're seeing is a qualitative influence factor at work,"* says Richard Heckel, president of Engineering Trends. He explains that

the recent trends are due in part to generational effects. "*Back in the old K through 12 system, girls weren't supposed to get involved in the quantitative stuff, the math and science,*" he says, "*and most engineering fields, traditionally, were seen as quantitative. All that has changed.*" Now disciplines like biomedical engineering are seen as more qualitative, something women may see as more accessible.

Many women grow up interested in engineering but want to do something that will affect people on a personal level. Biomedical engineering can have a direct impact on people's lives and on their health, as opposed to making a better gear for a car or designing it to go faster.

Other engineering fields that are attractive to women are:

- Environmental Engineering (40.6%)
- Agricultural Engineering (37.1%)
- Chemical Engineering (36.5%)
- Industrial/Manufacturing (34.6%)
- Architectural (31.9%)
- Materials/Metallurgical (31.2%).

*Percentages represent number of degrees awarded to women.*

# Team Buiding Activities

# THE WHALE BAND-AID®

## A Team Approach to a Large Problem

Time Required: 30-45 minutes to solve the problem
20-30 minutes for questions and discussion

## How this Learning Experience Meets the National Science Education Standards:

As a result of activities in grades 5-8, all students should develop:

### Content Standard A: Science as Inquiry

Abilities Necessary to Do Scientific Inquiry

- Develop descriptions, explanations, predictions, and models using evidence.
- Think critically and logically to make the relationships between evidence and explanations.
- Recognize and analyze alternative explanations and predictions.
- Communicate scientific procedures and explanations.
- Understand scientific inquiry.
- Different kinds of questions suggest different kinds of scientific investigations.

### Content Standard G: History and Nature of Science

Science as a Human Endeavor

- Women and men of various social and ethnic backgrounds engage in activities of science, engineering, and related fields; some scientists work in teams, and some work alone, but all communicate extensively with others.
- Science requires different abilities, depending on such factors as the field and study and the type of inquiry Nature of Science.
- Scientists formulate and test their explanations of nature using observation, experiments, and theoretical and mathematical models.
- It is part of the scientific inquiry to evaluate the results of scientific investigations, experiments, observations, theoretical models, and the explanations proposed by other scientists.

## Teacher Notes:

As the facilitator, you will determine what criteria will determine how successful the group is at the end of the activity—how long it takes them to complete the task, how many members of the group are "knocked off into the water" during the activity, if they develop a feasible plan, if they try several different strategies, if they get the covering turned over at all, etc.

Students can complete a journal entry where they describe what they think would be a successful plan for another group to follow. If time permits, and you feel it would be beneficial, have the class repeat the activity to see if their problem solving skills improved.

## Safety Notes:

Be certain that the covering that has been selected will actually allow the number of individuals involved to stand on it. Attention should be given to maintaining an area that is clear of anything that could pose a danger for those engaged in the experience (any obstacles in the area). Manage the whole group while the students solve the problem to be certain that no horseplay interferes with the process or compromises anyone's safety. Should the group include individuals with special needs, modifications of accommodations will need to be addressed.

## Getting Started:

1. Gather the materials needed for the experience.
2. Identify an area large enough for the whole group to safely engage in the activity; there should be no desks, chairs, tables, etc. in the area.
3. Copy student sheets.

## Materials Needed:

- An old mattress cover, sheet, blanket (any type of covering will do, and the size of it will be determined by the number of students involved; 20-22 students can work on a standard double mattress cover)
- An area large enough to spread the covering out flat on the floor and the entire group stand on it
- Student sheets

## Procedure:

1. Have the students assemble in the appropriate area for the activity.
2. Conduct the experience as follows:
   - With no previous information or directions, point out the covering on the floor.
   - Instruct the entire class to then gather and stand on the covering only—no part of anyone's body can be in contact with anything other than the covering and each other.
3. After the group is in place and still, give them the following information:
   "Close your eyes and listen very carefully to what you are about to be told. You are no longer with me at school. You and your classmates are now in the middle of the ocean. And you are standing on a giant Band-Aid® on a whale's back. The only problem is that the Band-Aid® is upside down! Your task is to come up with a way to turn the Band-Aid® over without anyone in the group stepping off the Band-Aid®. I will offer no assistance nor answer any questions. The group must develop a plan while standing together and working together."
4. Instruct the students to then begin solving the problem.
5. Following the completion of the experience, have the students complete the student sheet.
6. Facilitate a discussion that builds on the student responses.

# The Whale Band Aid
# Student Sheet

Following the group activity, answer the following questions while reflecting on your group's ability to successfully solve the Whale Band-Aid® problem.

1. What did you find to be the biggest problem in solving the problem with the Whale Band-Aid®?

2. What was the most frustrating part of the activity?

3. What was the most difficult part of the activity?

4. What skills do you think were necessary for the group to achieve success?

5. If you could change one thing that the group did during the activity, what would it be?

6. What do you think should have been the first thing the group did?

7. Did your group work as a team? How could this have been improved?

8. How might this activity be similar to doing a science lab or activity?

9. What did you learn about yourself during this activity?

# Carbon Copy Creations

## Cooperatively Copying an Original Creation

Time Required: 15-20 minutes for the activity
20-30 minutes for questions and discussion

### How this Learning Experience Meets the National Science Education Standards:

As a result of activities in grades 5-8, all students should develop:

### Content Standard A: Science as Inquiry

Abilities Necessary to Do Scientific Inquiry

- Use appropriate tools and techniques to gather, analyze, and interpret data.
- Develop descriptions, explanations, predictions, and models using evidence.
- Think critically and logically to make the relationships between evidence and explanations.
- Recognize and analyze alternative explanations and predictions.
- Communicate scientific procedures and explanations.
- Use of mathematics in all aspects of scientific inquiry.
- Understand scientific inquiry.
- Different kinds of questions suggest different kinds of scientific investigations.
- Technology used to gather data enhances accuracy and allows scientists to analyze and quantify results of investigations.
- Scientific explanations emphasize evidence, have logically consistent arguments, and use scientific principles, models, and theories.

### Content Standard E: Science and Technology

Abilities of Technological Design:

- Design a solution or product.
- Implement a proposed design.
- Evaluate completed technological designs or products.
- Communicate the process of technological design.

### Understanding Science and Technology

- Scientific inquiry and technological design have similarities and differences.
- Technological designs have constraints.
- Technological solutions have intended benefits and unintended consequences.

## Content Standard G: History and Nature of Science

### Science as a Human Endeavor

- Women and men of various social and ethnic backgrounds engage in activities of science, engineering, and related fields; some scientists work in teams, and some work alone, but all communicate extensively with others.
- Science requires different abilities, depending on such factors as the field and study and the type of inquiry.

### Nature of Science

- Scientists formulate and test their explanations of nature using observation, experiments, and theoretical and mathematical models.
- It is part of the scientific inquiry to evaluate the results of scientific investigations, experiments, observations, theoretical models, and the explanations proposed by other scientists.

## Getting Started:

1. Obtain enough LEGO blocks for each group to have about 20; each group's set MUST be identical same number, type, and color of blocks.
2. Using the same set of blocks as each cooperative set, build an original structure; set the structure inside a box so that it is not visible to the whole group. (this can be ANY type of structure)
3. Using the template provided, prepare group sets of task cards by placing each set in a zip lock bag.
4. Copy student sheets.

## Materials Needed per Group of Students:**

- One zip lock bag containing LEGO building blocks (about 20 per bag)
- One set of group task cards
- Student sheets

**Please note that when assembling the bags of LEGO blocks, each bag MUST contain an identical set of blocks: same number, type, and color of blocks. THIS IS ESSENTIAL!

## Procedure:

1. Assemble the students into cooperative groups of four students each (group size may be modified based on class size, student ability, or teacher objectives.)
2. Explain the following task:
   - Their job is to exactly duplicate the structure that the teacher has built.
   - Each group is to use their bag of LEGO blocks, which is identical to the set that the teacher used.
   - The structure that they will attempt to copy is in a box hidden from them.
3. Give each group a set of task cards; the groups will assign the following tasks by passing out the cards within the group: "builder", "manager", observer", and "active listener".
4. Explain the group job as follows: The job of the observer is to go look in the box where the teacher's structure is, observe the structure for one minute without talking, then report back to the group. The observer MAY NOT handle the teacher's structure! In fact, the observer may not use hands at all - hands must remain behind the back or in pockets once rejoining the group. Upon returning to the group, the observer will instruct the builder on how to build the structure. The builder has two minutes to assemble the structure as directed but cannot ask questions or make comments. The observer can only give VERBAL instructions - no gestures, pointing, or touching. Only the builder can handle the LEGO blocks. The manager makes sure the observer keeps his/her hands behind the back. The manager will also remind the builder not to ask questions or comment. The active listener will ask questions to clarify its meaning. When the teacher has called time at the end of the two-minute period, cards are passed in a clockwise direction and each member of the group then takes on a new role. The procedure is repeated until all members of the group have had the opportunity to function in each role. It is not necessary that the structure be disassembled after each rotation - the builder must simply follow the observer's instructions.
5. After the group members have performed each of the four roles, the group is then allowed to choose the person they think is best at each job, the students will then assume that role and the group has one final opportunity to complete the task of replicating the teacher's LEGO structure.

6. Following the completion of the rotations, have the students complete the worksheet provided for reflection.
7. Facilitate a discussion with the students about the importance of individual jobs within the cooperative groups: emphasize the need for teamwork and how each individual has strengths and weaknesses that can be used to make the team more successful.
8. After engaging the students in the whole group discussion, inform the students that the procedure will be repeated one more time, but with this one important difference - when the students observe the second teacher-built structure, they will be allowed to take notes as they observe.
9. After they have completed the procedure and constructed their second structure, use this experience in the discussion to emphasize the importance of a well-kept notebook.

# GROUP TASK CARDS

| | |
|---|---|
| BUILDER | OBSERVER |
| ACTIVE LISTENER | MANAGER |

# Carbon Copy Creations

## Questions and Answers For Group Reflection

## Student Sheet

1. What skills were necessary for success in this activity?

2. What were the greatest frustrations in each of the following jobs?

3. Why was teamwork necessary for success in achieving your objective?

4. How is this activity similar to doing a science or math learning experience? How is it different?

5. What would you do differently if you repeated this activity with a different structure?

6. What skills are necessary for one to achieve success in math or science?

LARGE TASK CARDS

| | |
|---|---|
| **BUILDER**<br>The only person who can touch the LEGOs; IS NOT ALLOWED TO SPEAK; must follow the observer's verbal instructions only | **OBSERVER**<br>The only person who can give instructions to the builder; actually sees the structure to be copied and relays information; MUST KEEP HANDS BEHIND THEIR BACK AT ALL TIMES |
| **ACTIVE LISTENER**<br>Will ask questions to help clarify the observer's instructions; may rephrase what the observer has said in order to clarify its meaning; MAY NOT USE THEIR HANDS | **MANAGER**<br>Makes sure that no one touches the LEGO except the builder; makes sure that the observer keeps their hands behind their back; makes sure the builder asks no questions or makes any comments |

# Carbon Copy Creations
## Teacher Sheet with Suggested Answers

1. What skills were necessary for success in this activity?
   ***Good observation skills, good memory, ability to listen, ability to communicate clearly, teamwork, patience, ability to follow directions***

2. What were the greatest frustrations in each of the following jobs?
   ***a. Builder - Not understanding what the observer was talking about, not being able to speak or ask questions, not having a mental image of the structure, time limitation on doing what you're told to do***
   ***b. Manager - Making sure the observer doesn't use their hands or touch the structure, keeping the builder from talking to the observer, finding the right way to clarify what is being asked or said, not being able to touch the structure***
   ***c. Observer - Not being able to use hands or touch the structures, unable to keep a clear image in their mind, time limitation on observations, the description you give doesn't match what the builder does, can't take notes during observations***
   ***d. Active Listener - Finding a way to correctly interpret what the observer is saying so that the builder understands it, not being able to touch the structure***

3. Why was teamwork necessary for success in achieving your objective?
   ***The job was too complicated for a single individual to accomplish. People have skills in different areas—each role required different skills. The structure of the task required teamwork.***

4. How is this activity similar to doing a science or math learning experience? How is it different?
   They're s***imilar in that you are making observations, looking for patterns, applying that information toward an end product (in this case, constructing a model), working as a team, coming to consensus, assessing individual strengths and abilities, keeping an accurate written record, and the need to communicate.***

   ***They're*** d***ifferent in that you didn't come up with your own question or problem and therefore developed no hypothesis to test, not data collected, no mathematics integrated, no plan developed from the beginning, and there was no inquiry.***

5. What would you do differently if you repeated this activity with a different structure?
   ***Develop a logical plan for making observations, take written notes, if possible. Consider whether or not the structure should stay in the same orientation and not be moved around, develop a common vocabulary to describe the geometry.***

6. What skills are necessary for one to achieve success in math or science?
   ***Observation skills, communication skills, teamwork, recognizing patterns, developing a plan, interpreting data, drawing logical inferences from the data.***

# CREATING A COLLABORATIVE CLAN

Time Required: Up to 45 minutes (including presentations)

## A Learning Experience that Facilitates the Transition from a Whole Group to Cooperative Groups

### Getting Started:

1. Make sets of cards that equal the number of cooperative groups within the whole group. Within the card sets, make certain the number of cards is equal to the number of individuals in the group. The total number of cards should equal the number of people in the whole group.
2. Prepare a large copy of each type of different cards used in the card sets prepared and place one at each group work station.

### Materials Needed:

- Sets of small cards
- A large version of each different picture featured on the small cards
- Adequate space for free movement of the whole group
- Work stations or meeting spaces for cooperative groups
- Student sheets

### Procedure:

1. Participants will begin the learning experience in a whole group setting before they move into smaller groups according to the following guidelines:
2. Pass out the small cards, face down, and when all members of the whole group have a small card, begin the experience.
   - When each individual receives a small card, they will then find the other group members who have the same card.
   - After finding the matching small group members, the groups will then find the large version of their small cards and assemble in that work area.
   - Group members then become acquainted with one another quickly and prepare for the phase that follows.

### Safety Notes:

The only safety concern would be for the mobility of all participants. Should the group include individuals with special needs, modifications of accommodations will need to be addressed.

# COLLABORATIVE CONSENSUS (part 2)

Emphasizing cooperative learning and consensus building while facilitating communications skills.

## Prior to the Activity:

Classroom management is improved if basic supplies needed by collaborative groups are put into containers that can be easily accessed by each group. Tupperware boxes can be stocked with markers, crayons, tape, rulers, glue, scissors, calculators, pencils, and other common materials. Preparing these before beginning in collaborative group work can be a true asset to classroom management techniques. If materials are not already set up in group containers, gather markers for each collaborative group, and be certain that you have newsprint or poster board for each group as well. Determine how much time will be allotted for the completion of this experience and have students begin in their cooperative groups.

## Safety Notes:

Adequate space for the collaborative groups to engage in this experience is the main safety concern. Table tops should be constructed so that groups can safely assemble together and complete the task assigned without difficulty.

## Getting Ready:

1. Establish effective classroom management techniques.
2. Assemble general group materials such as markers, scissors, tape, glue, etc. into group containers.
3. Obtain other materials needed.
4. Copy student sheets.
5. Conduct "Creating a Collaborative Clan" experience.

## Materials Needed:

- Student sheets
- "Creating a Collaborative Clan" group card
- Cooperative group work area
- Newsprint or poster board
- Markers, crayons, pencils, pens, etc.

## Procedure:

1. Collaborative groups will use their large group card from the "Creating a Collaborative Clan" experience to do the following:
   - Use the card as the focus when answering the questions listed on the student sheet.
   - Make certain that all answers relate in some way to the picture on the group card.
   - The members of the group must reach consensus when determining the answer to each question.
   - Prepare a poster presentation on the newsprint (or poster board) that visually summarizes the information given in the group responses.
2. Each group will conduct a poster presentation and share their responses as well as the rationale behind each.
3. Post all posters in the learning environment.

# COLLABORATIVE CONSENSUS
## Student Sheet

Using your group cards as your focus, have your team come to a consensus and have a group response ready for each of the following questions. Remember that all answers must in some way relate to the picture on your card!

1. What is your team name?

2. What will your team's motto or slogan be?

3. Where would your team prefer to eat?

4. What is your team's song?

5. What famous person would most likely join your team and why?

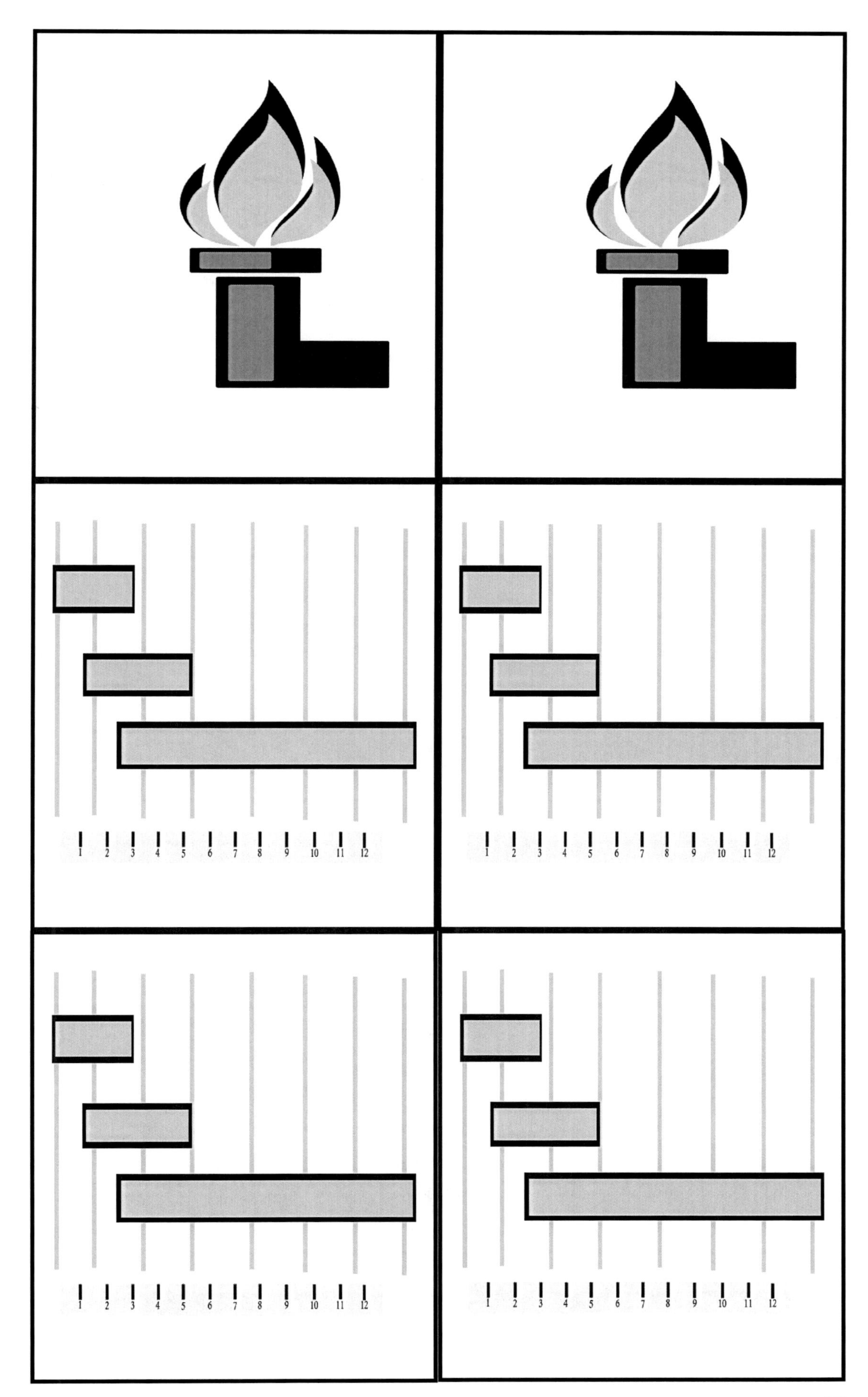
1 2 3 4 5 6 7 8 9 10 11 12
1 2 3 4 5 6 7 8 9 10 11 12
1 2 3 4 5 6 7 8 9 10 11 12
1 2 3 4 5 6 7 8 9 10 11 12

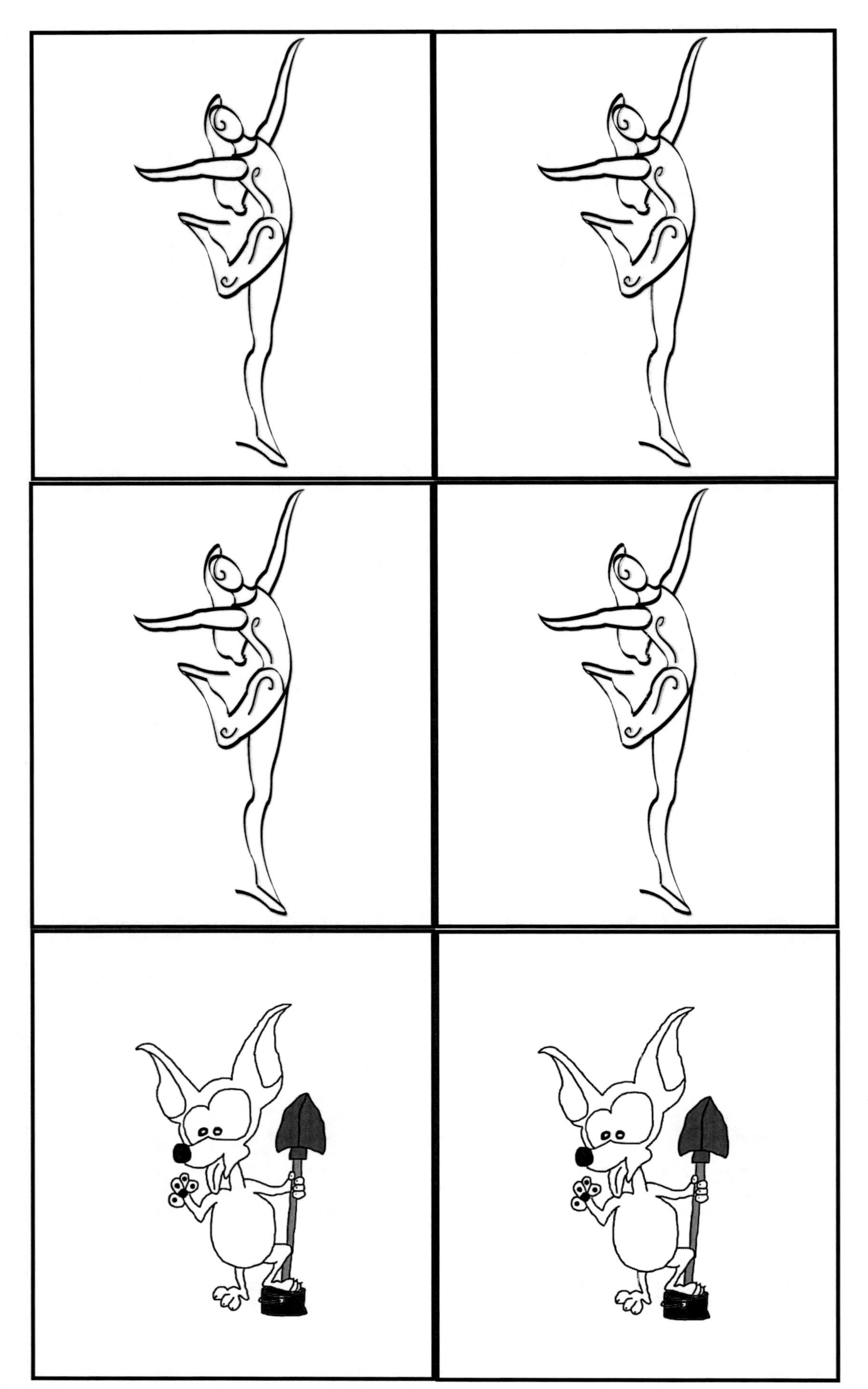

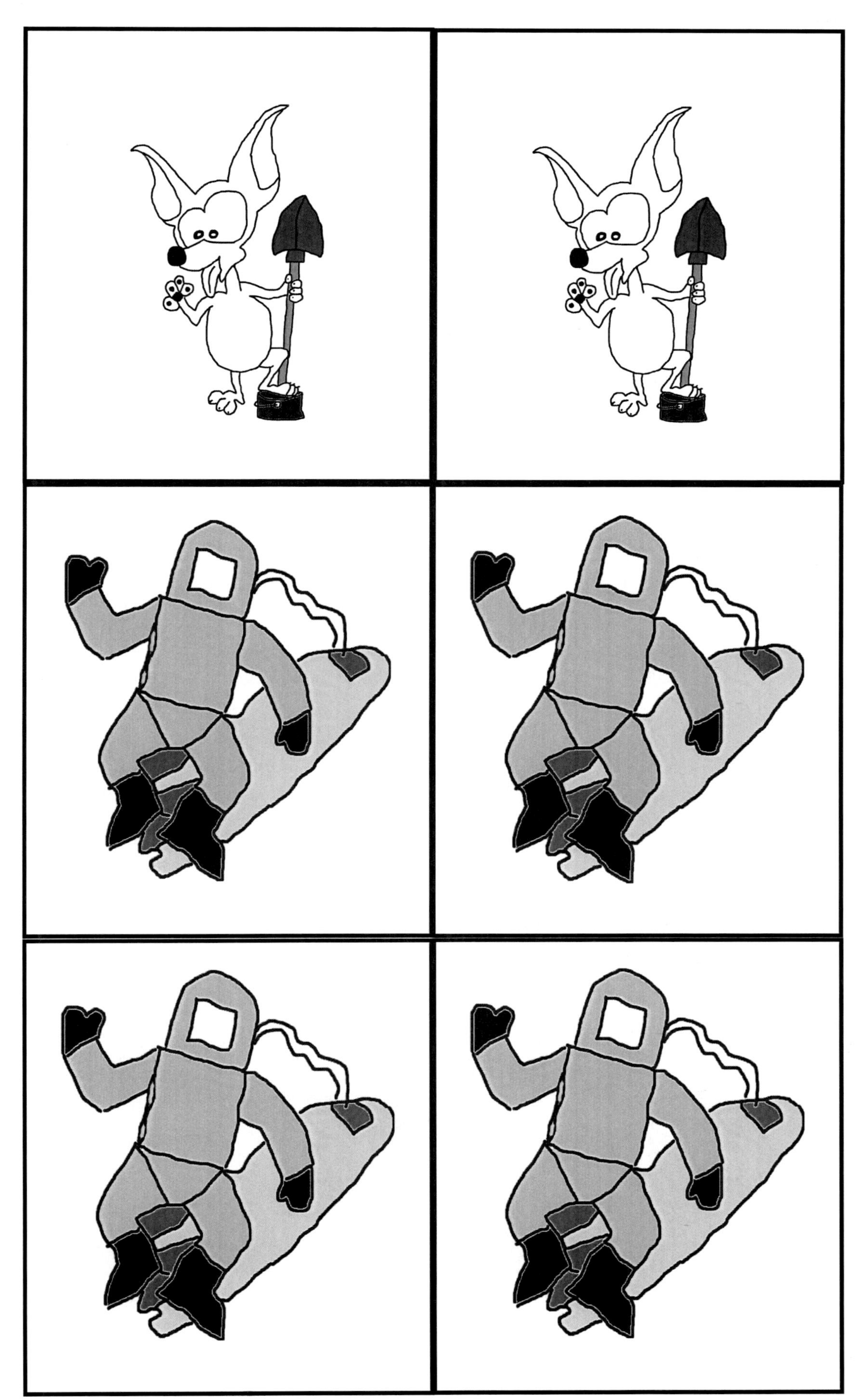

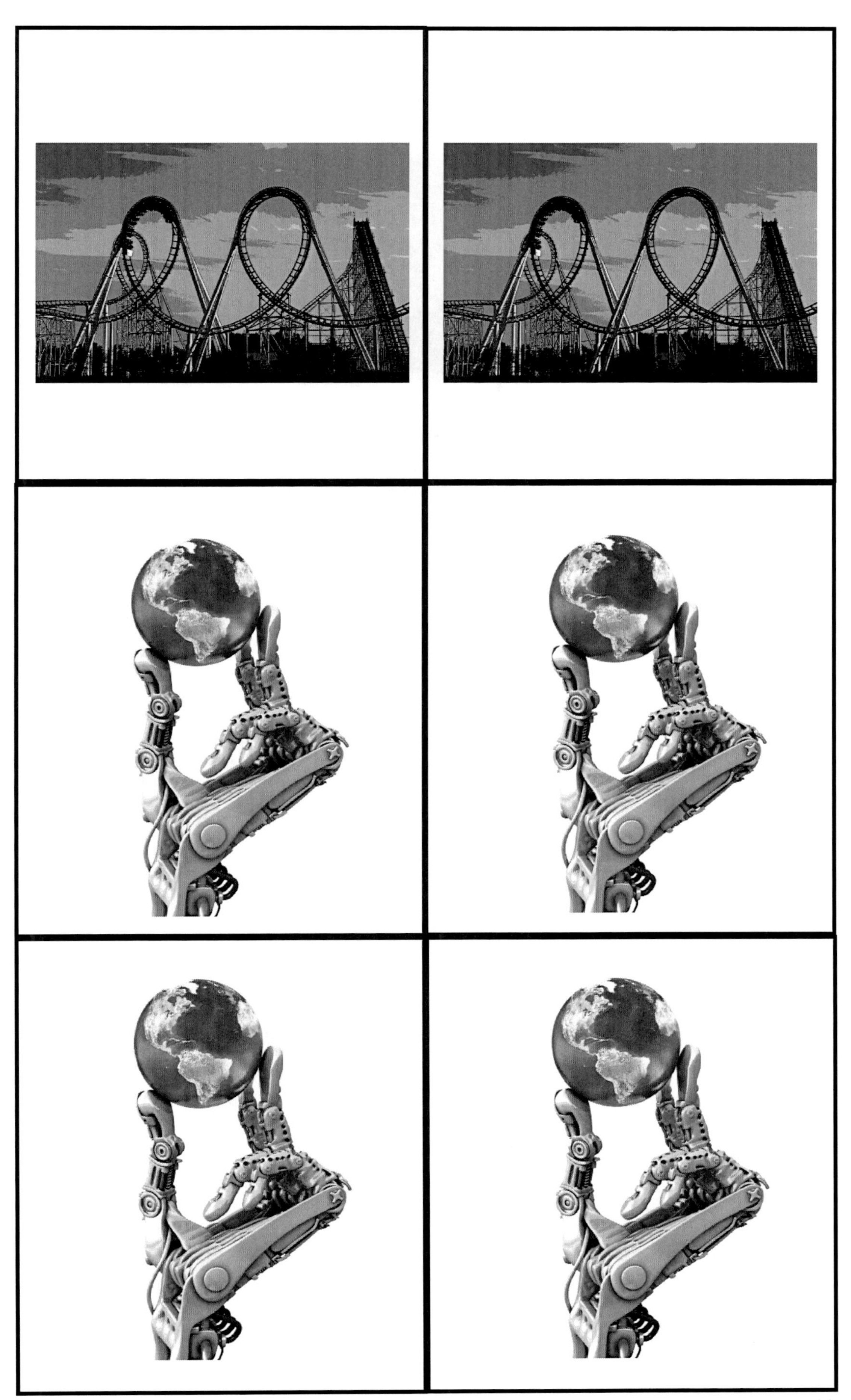

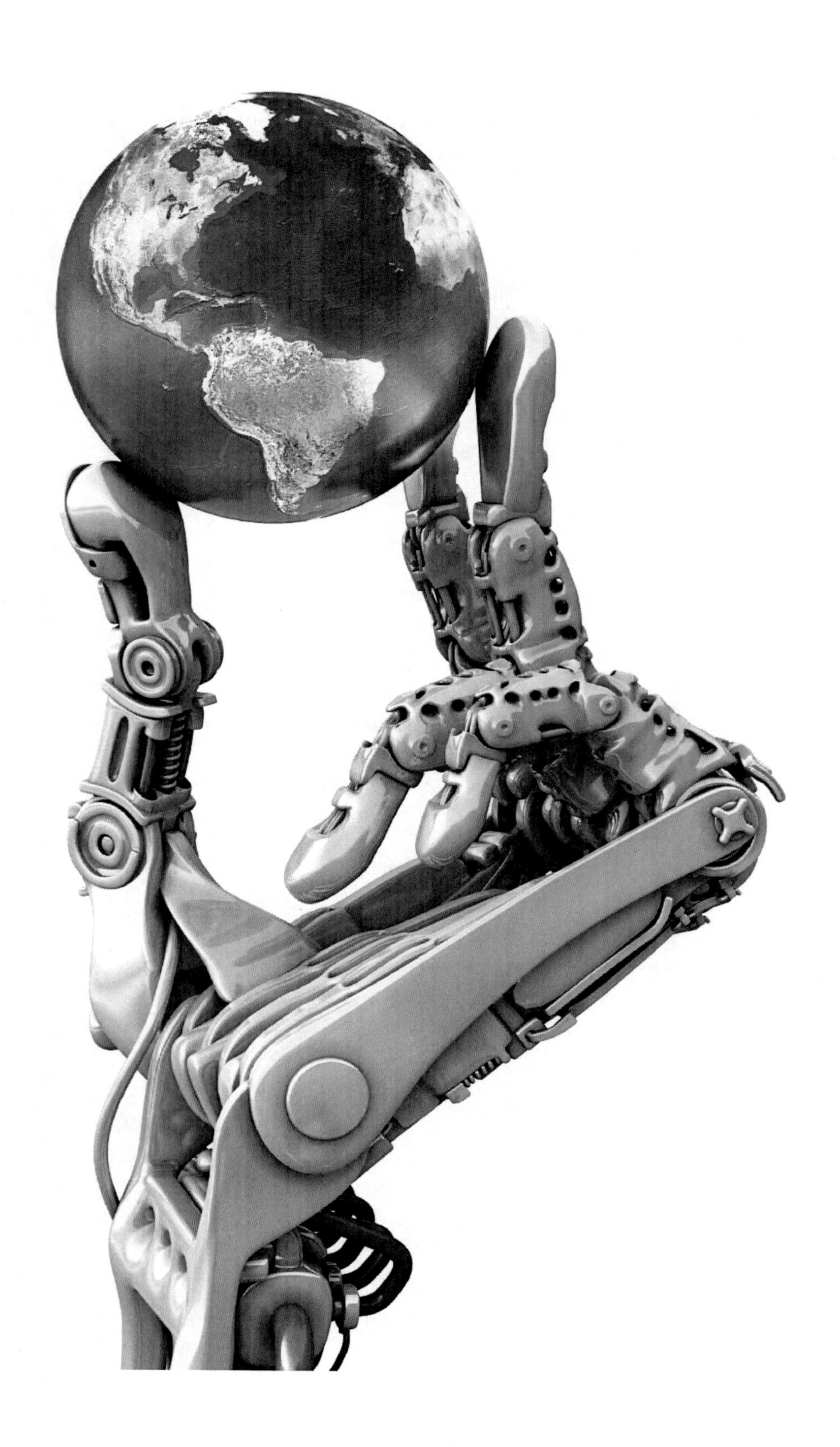

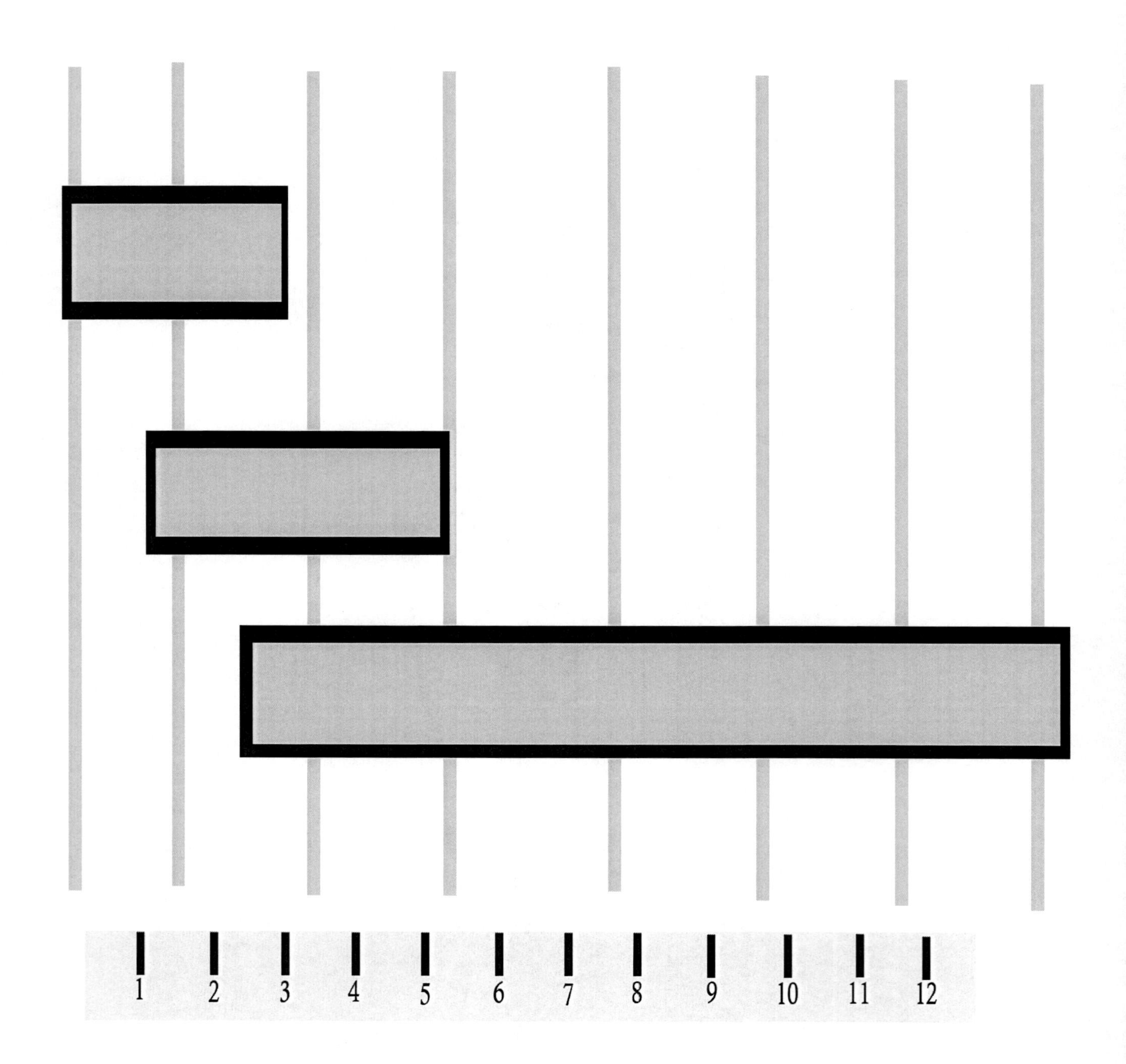
1
2
3
4
5
6
7
8
9
10
11
12

# What's in the Bag?

Time Required: 20 minutes for the activity
20 minutes for sharing

## Developing Rules for Cooperative Groups to Live by

## Teacher Notes:

The activity can be conducted in a variety of ways.

- The student sheet provided can be used for the groups to record their different ideas as well as their final responses for each item. When reporting their ideas, the teacher may want to have the students engage in different strategies as they share.
- Each group can be given a different colored set of sticky notes to write their rules on. They can then participate in a carousel type activity that allows the groups to rotate through six different stations, one for each item in the bag.
- Before beginning the carousel, a member of the cooperative group will place a sticky note with their group's rule for that item on a poster board/newsprint/butcher paper at each station. When each group has placed their six sticky notes at the correct stations, the students will then engage in a carousel where they visit each station and review the different responses given. While at each station, the groups will determine which rule they feel is best for the item presented and record why they chose it.
- When the carousel is completed, the teacher will then lead the class in a discussion of the different responses given and the best ones selected.

## How to Do a Carousel Activity:

- Divide the class into cooperative groups.
- Assign areas of the room that will serve as "stations" for the activity.
- At each station there should be a task, question, or reading to be completed.
- Explain the following concept to the students. They will have a specific amount of time to complete the task that is required at each station. While at the station, they must work as a group and remain there until the signal to rotate is given. They are not to move ahead or try to work ahead during the activity. If they do not finish their task within the time allotted, the groups must still rotate to the next station when told to do so. Be focused and work together at each station!

- At the end of the rotation series, each group will have completed the entire assignment given for the carousel activity.
- Walk through the rotation if necessary to insure that the students are aware of the direction they are to rotate.
- Ask the students if they have any questions concerning the procedure or activity.
- Announce how much time they will have at each station.
- Have each of the groups go on to a different station, staggering the groups if possible.
- Begin the activity and monitor the groups as time is kept.
- When the rotations are complete, have the groups return to the designated area for follow-up and discussion.

## Getting Started:

1. Obtain materials
2. Purchase zip-lock bags
3. Prepare group bags with one of each item in each bag
4. Copy Student sheets

## Materials Needed Per Group of Students:

- One zip-lock bag containing one tootsie pop, a paper clip, a penny, a rubber band, a toothpick, and a poker chip
- Student sheets
- Poster board, newsprint or butcher paper
- Six sticky note pads

## Procedure:

1. Assemble students into cooperative groups of four students each (group size may be modified based on class size, student ability, or teacher objectives.)
2. Give each cooperative group a prepared bag.
3. Have the members of each group brainstorm ideas for what each item could represent as a "rule" for cooperative groups to use when working together; emphasize the need to use the characteristics of the item or what it is used for. (For example, the rubber band = "be flexible".)
4. Each rule should be something that will help the groups to work together efficiently, cooperatively, productively, and responsibly (These are also good rules to follow in everyday life!) Each group must come to a consensus as to what the rules for each respective item will be.
5. Give the groups a time limit to come up with their list of rules, and then have the class as a whole share their ideas; see how many were duplicated and what the most original rules were.
6. Facilitate a good discussion of how cooperative groups should work together as well as the importance of critical thinking and good communication and socialization skills.

# What's in the Bag?
## Student Sheet

Record all group ideas for the rules developed for each of the following items. Indicate your final response in the space provided.

1. Rubber Band

   Rule:________________________________________________

2. Penny

   Rule:________________________________________________

3. Toothpick

   Rule:________________________________________________

4. Paper Clip

   Rule:________________________________________________

5. Poker Chip

   Rule:________________________________________________

6. Tootsie Pop

   Rule:________________________________________________

# What's in the Bag?
## Carousel Sheet

As you carousel through the activity, record the different ideas presented by the class groups. Indicate which ideas were duplicated and how many times. Through your group's discussion, determine which "rule" you think should be established for each item. Underline each final response on your sheet.

Station #1 Item:________________________________________

Station #2 Item:________________________________________

Station #3 Item:________________________________________

Station #4 Item:________________________________________

Station #5 Item:________________________________________

Station #6 Item:________________________________________

# HAMPERED BY HEIGHT

Time Required: 20-30 minutes.

## Promoting Problem Solving Skills while Emphasizing Teamwork

## Teacher Notes:

### Solution to the Problem:

In determining a way to average individual heights without actually engaging in mathematical computation, each group's procedure must collectively represent the total height of the group members and then divide by the number of people involved. Most solutions involve the following process:

1. Unroll the adding machine tape and beginning at the end, stretch the tape the length of the first member of the group. Do not cut.
2. Mark the tape with the pencil to show the height of the first individual.
3. Beginning with the first mark, now measure the second individual. Use the first mark to begin at their feet and stretch the tape to the top of their head. Mark their height. The tape should now show a representation of two group members height.
4. Follow this procedure until all group members have recorded their height on the adding machine tape. The final mark should provide an indication of how long the tape should be to include every group members height. (Think of all group members lying on the floor head to toe and then stretching the tape along beside them—it should reflect the same length.)
5. Cut the adding machine tape at the final mark.
6. If there are four group members, fold the tape over to form two halves; this represents dividing by two, although no actual division has taken place. Follow this step by then folding again to divide the tape now into four equal pieces; this represents dividing by four and should simulate the mathematical process for averaging.
7. If there are three group members, fold the tape until it appears that you have three equal pieces represented. Overlap the tape until the three equal portions are evident.

8. Using the meter stick or measuring tape, measure one of the equal pieces of adding machine tape. Record the measurement in centimeters.

9. Have the groups report back with their solutions.

10. Make certain that the group members can justify the procedure they engaged in to accomplish this task. In leading the discussion, emphasize how you must think about what it takes to actually obtain an average and how you can do that without using mathematics.

## Safety Notes:

Exercise caution in working with scissors.

## Getting Started:

1. Gather the materials for the experience.
2. Determine how many students will be in each team.
3. Identify an appropriate area where each team can spread out to work on the problem without interfering with the work of other teams.

## Materials Needed Per Group of Students:

- One roll of adding machine tape
- Scissors
- Pencil
- Meter Stick or Measuring Tape

## Procedure:

1. Divide the class into cooperative groups of four students each (group size may be modified based on class size, student abliity, or teacher objectives.)
2. Instruct them that they are to complete a measurement task following a specific set of rules: if they do not adhere to the rules as stated, their response will be rejected.
3. Provide the objective as follows: Each group is to determine the average height of the members of the group using only the materials provided and following these rules:
   - You can only use the scissors to cut once.
   - You can only use the meter stick to measure once.
   - You must measure in centimeters.
   - You cannot add, subtract, multiply, or divide at any time.
4. Be certain that the students understand that there can be NO mathematical computations done at any time. No calculators are to be used.
5. The meter stick or measuring tape can be used only once, making one complete measurement. This means that if the students need to turn the stick or tape end-over-end, that still counts as being used only once.
6. Make certain that the final response is presented in centimeters, and that the group can explain or justify their solution to the problem.

# HAMPERED BY HEIGHT

## Cooperative Group Challenge Sheet

Determine the average height of the members of the group using only the materials provided and following these rules:

- You can only use the scissors to cut once

- You can only use the meter stick to measure once to make one complete measurement; that means if you need to turn the measuring instrument end over end, that still counts as being used only once

- You must measure in centimeters

- You cannot add, subtract, multiply, or divide at any time; there can be NO mathematical computations done at any time

- No calculators are to be used

# Chemical Engineering

### Average Starting Salary Right Out of College: $52,000

As creative and innovative problem solvers, chemical engineers enjoy great diversity in their intellectually challenging field. Chemical engineering offers one of the highest starting salaries of all the engineering disciplines. Everything that our senses enjoy consists of chemicals in one way or another. Chemical engineers have worked on creating the purple rose that has no thorns, the caramel on a caramel apple, and even your tennis shoes. The chemical engineering profession has improved water and waste systems, created new drugs and drug delivery systems, and improved the crop yields for farmers. Most chemical engineers work in manufacturing, pharmaceuticals, health care, design and construction, pulp and paper, petrochemicals, food processing, specialty chemicals, microelectronics, electronic and advanced materials, polymers, business services, biotechnology, and the environmental health and safety industries.

Chemical engineers can choose from many specialties within the discipline. A chemical engineering student who picks a track in environmental engineering might be interested in reducing pollution or producing better food. Chemical engineers who focus on biomedical engineering are often called biochemical engineers; they may design new or improved artificial organs. If you are more attracted to the big picture, you might see yourself looking for ways to streamline processes or increase safety with a specialty as a process design engineer.

Traditionally, pharmaceutical, petroleum and chemical companies employed the bulk of the chemical engineering profession. Pharmaceutical companies still employ a large number of chemical engineers to research, develop, or design their product lines. Now, however, many chemical engineers work in biotechnology, material science (such as the plastics, rubber, ceramics, and metals industries), and electronics. The food industry, the Department of Energy, and the Environmental Protection Agency have also become prominent employers of chemical engineers.

Many chemicals engineers have opted for careers in research. A chemical engineer who goes to medical school could become a medical doctor specializing in rare diseases or a medical researcher. A chemical engineer who goes to law school could be a patent attorney or a specialized attorney for a company that has invented a new drug or drug delivery system.

## Shoes

Jennifer Ocif,
Performance
Engineer, Reebok
International, Ltd.

In 1993 with an undergraduate degree in mechanical engineering from Rensselaer University, Jennifer began working at a medical device manufacturer in the research and development department. Jennifer worked with biomedical engineers who had interesting stories about their course work and piqued her interest to go back to graduate school. She wanted to study something that would become the foundation for a new, people-oriented and human-applied career that would be interesting, challenging and fun. She had always been an athlete and was curious to apply engineering principles to the human body in motion. That's when she decided she wanted to work as an engineer in the sporting goods industry.

In 1995, while working towards a Masters in biomedical engineering at the University of Iowa (UI), Jennifer heard about a design contest that was being sponsored by the Sport Science and Technology Division of the United States Olympic Committee (USOC) in Colorado Springs, Colorado. It sounded so cool to design something that would help US Olympic athletes win gold medals. The 1996 Summer Olympic Games were coming to Atlanta the following year, so USA Olympic spirit was in high gear. Despite her heavy graduate school workload, she entered the contest. It was the contest of a lifetime because she was the perfect contestant; an engineer studying sports biomechanics. She knew she would have fun tinkering with a homemade science project to see what she could come up with to help US athletes.

She consulted with Dr. Jim Hay, her advisor at UI and one of the top sport biomechanists in the world studying track and field. Based on his research, she developed an inexpensive motion-tracking device to pinpoint the location of an athlete's footfalls on the track during training or competition. Dr. Hay's research had shown that foot placement, especially during the approach of a triple jump or long jump, directly affects an athlete's performance. At the time, Dr. Hay's research with track and field athletes depended on an expensive video motion analysis system that required several months of data analysis after the day of filming. Frustrated athletes needed performance feedback faster. Dr. Hay challenged Jennifer to build a small, lightweight, inexpensive, real-time data acquisition and analysis motion tracking system that even high school, college and Olympic coaches could use to help train their "Gold medal" athletes. Dr. Hay wanted to enable coaches to see an athlete's foot placement relative to the take-off board before the athlete got up from the sand pit after completing the triple jump or long jump.

Jennifer investigated possible solutions and submitted a design proposal to the USOC judges who were "real" sports biomechanists who had been working with Olympic athletes for years. A few months later, she received a letter that her design had been chosen as one of five finalists out of more than 20 applications! Excitedly, she began to build a prototype using ultrasonic transmitters, ultrasonic receivers, parts bought at Radio Shack, plastic orange cones and wires. She used her electrical engineering textbooks to design the circuitry and consulted an electrical engineer to help build the circuit boards. She tested the assembled prototype by attaching it to the laces of her own sneakers and ran at the university track. Jennifer's device resembled a car key-chain remote control in both weight and size. It took about 6 - 12 months to develop and $500 to build the first working prototype after learning from a few failures. Ironically, those early failures were key to her success!

In 1996, she traveled to the US Olympic Training Center in Colorado Springs with Dr. Hay to present her final design and prototype to the USOC judges. The competition was tough, but the judges awarded her a first-place prize of $1,000, a cool trophy and an exclusive tour of the sports biomechanics labs where she saw several Olympic athletes in training. From this point on, she knew that she would combine sports and engineering into a fun, intellectually and monetarily rewarding career. She hopes to one day help USA Olympic and Paralympic athletes win gold medals.

At Reebok, Jennifer is responsible for designing, engineering, developing and evaluating athletic footwear with a team of coworkers in the US and in Far East countries with the goal of marketing and selling performance-driven footwear for athletes both young and old, male and female, recreational or competitive, worldwide.

## Shoe Design Today

Running, biking, climbing, hiking, playing basketball, football, soccer, skateboarding, speed skating, rollerblading, tennis, volleyball, bowling, and many more sports all have two things in common. Shoes and the engineers that design them! Almost every sport has shoe requirements. It takes engineers to analyze the movements and cushioning needs of the sport to determine the best shoe possible, — One that will reduce injuries, enable outstanding performance, be comfortable and look great.

The Olympic contenders of the early 1900's had shoes made of leather that probably weighed two or three pounds. The Olympic contenders of today were shoes that are only two or three ounces. Shoe designers must accommodate the requirements of the sport for which they are designing.

- A skateboarder wants a shoe with a sticky sole that will grip the board and provide more traction.
- A bicycle rider wants a shoe where the sole does not flex so that more power can be transferred to the pedal.
- A basketball player wants a shoe that will grip well when they are running but not when they are turning or pivoting.
- A runner wants a shoe with good cushioning that won't break down quickly.
- A marathon runner wants a shoe with built-in microchips so they can know their exact position and time their runs more closely.

According to Dolores Thompson, a process engineer at Nike, "It has been helpful to have a familiarity of sports, fitness and teamwork. Specifically, to understand the different performance needs of cushioning. Cleated footwear (soccer, football, softball) has different cushioning performance needs than court (tennis, basketball) or even running or rock climbing." To assess cushioning, designers have used everything from air pumps to silicon to other synthetics in an attempt to increase the fun and comfort of athletes everywhere.

Knees and Achilles' tendons are the most prominent injury areas sustained by athletes. In the Nike Sports Research Lab, scientists and researchers study the motion of athletes doing many different sports to design shoes that will help performance, be comfortable and minimize the risk of injury. To design a shoe that gives swifter kicks for soccer players, a high-speed video camera captures 1,000 frames per second of a player kicking the soccer ball. To design basketball shoes, researchers identify many different foot movements of players, analyze the motion with high-speed motion analyzers, measure the forces applied to the ground and record the pressure from sensors inside the shoe. Each and every movement must be recorded and analyzed to extend the greatest flexibility and performance potential to the athlete. The video helps the designers understand the dynamics of the sport, where the most pressure is applied during movement and the needs of the athlete in order to anticipate the requirements of the athletic shoe. Prototypes are then made and tested.

In addition to biomechanical applications of shoe design and analysis such as motion analysis, shoe companies also hire industrial, manufacturing and mechanical engineers to ensure the shoe manufacturing processes go smoothly. Materials have to arrive on time, inventory has to be controlled and the manufacturing processes have to be done according to specification. Dan Barch, an industrial engineer for Nike said, "The entire 'supply chain' needs to connect, like an organism — for real success. It all has to work together. Nike uses a lot of sport analogies, and this could be like basketball or soccer — you have your position and function, but can see everyone else and where the ball is. Sometimes you shoot, sometimes pass off quickly and sometimes hold. It depends on the situation and everybody's position. Any professional acting independently of the rest of the team — or team acting independently of the company — quickly finds themselves benched for the big plays."

Chemical and materials engineers will also find a wealth of employment in the shoe industry. Constantly on the lookout for new materials for soles and outer coverings, shoe manufacturers are in constant competition for the best cushioning, lightest overall design, most comfortable and best traction products. Finding new materials that add breathability for the long distance runner, springiness for basketball players, increased traction for skateboarders, flexibility and grip for wrestlers, more cushioning for long jumpers and/or strength and comfort for skeleton racers can make the shoe industry a challenging and rewarding field for an athletically minded engineer.

## Chemical Engineers may design:

### Agricultural Systems

Chemical engineers may help ease world hunger by designing better ways to produce food, and environmentally safer pesticides and fertilizers.

### Athletic shoes

Chemical engineers may work to develop or design new soles, fabrics or other materials for shoes.

### Baseball and football equipment

Chemical engineers may work to develop or design new light-weight materials for headgear that will be more comfortable and withstand greater impacts or forces. May develop new pills and windings for the insides of baseballs or covers and strings for the outside.

### Bowling

Chemical engineers may design balls, pins, synthetic lanes, ball polish, ball bags and oil for lane conditioning.

### Efficient Transportation

Chemical engineers may create cleaner fuels to power cars and other vehicles.

### Fishing rods and reels/equipment

Chemical engineers may work to develop or design new rods, reels, reel seats, grips, guide sets or any materials that will be stronger, make fishing more fun, or increase the sensitivity to fish bites.

### Helmets

Chemical engineers may work to develop or design new light-weight materials that will be more comfortable and withstand greater impact or force.

### Pharmaceutical Advancements

Chemical engineers may streamline industrial processes to produce life-saving drugs, vaccines, and antibiotics more efficiently and at lower cost.

### Pools and equipment for swimmers

Chemical engineers may develop fins, suits and/or performance training equipment. May develop new diving boards that allow more or less elasticity while maintaining their strength and durability.

### Skis and snowboards

Chemical engineers may work to develop or design new light-weight materials that will increase performance, be more comfortable and withstand greater impact or force.

**What's the difference between a chemist and a chemical engineer?**

Chemical engineers make about $20,000 more per year. Chemists work in a lab doing research. Chemical engineers also apply the results of chemical research and discovery in practical ways.

# The Puzzle of Nine

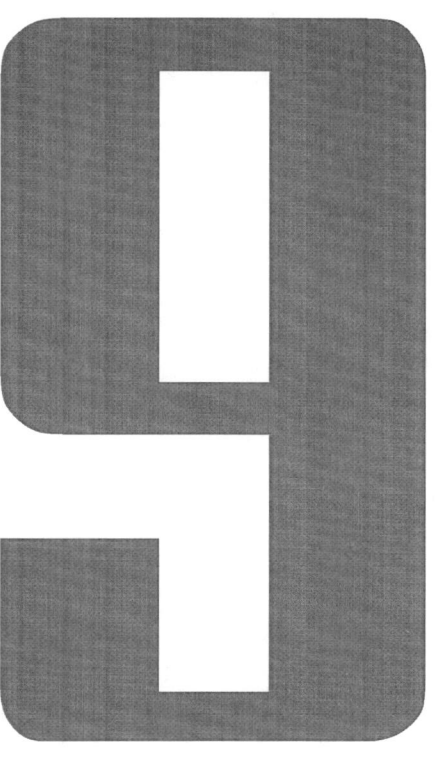

## Pondering a Puzzling Procedure

Time Required: 20 minutes

## The Problem:

Nine objects, identical in appearance, are placed on the bench. Eight of the objects have the same mass, while one has a slightly greater mass than the other eight. The only other item present is a double pan balance. Members of the audience are challenged to design a procedure to identify the object with greater mass that requires using the balance the fewest number of times.

## Getting Started:

1. Gather an adequate number of film canisters with caps.
2. Obtain the remaining materials needed for the challenge.
3. Copy student sheets.

## Materials Needed:

- Fifteen or twenty film canisters with caps
- A double pan balance
- Water
- Dropper or disposable pipette
- Student sheets

## Procedure:

1. Use the double pan balance to select nine film canisters that have the same mass.
2. Place several drops of water into one canister to give it about 10 percent more mass.
3. Place the nine canisters and balance before the class.
4. Present the following challenge (you may wish to make this a small group assignment): design a procedure to identify the object with greater mass that requires using the balance the fewest number of times.
5. After allowing time to develop a procedure, ask who can solve the puzzle using the balance four times; if someone volunteers, ask him or her to talk you through the procedure or demonstrate it to the class.
6. Ask who can solve the puzzle in three uses of the balance and demonstrate that procedure.
7. Ask who can solve the puzzle using the balance only twice; demonstrate the procedure.

## Teacher Notes:

## Using the balance three times:

- Place four canisters on each pan of the balance. If both groups of four balance, the heavy canister is the one still on the bench. If one group of four is heavier, it contains the heavy canister. Set the remaining five canisters aside.
- Place two canisters from the remaining group of four on the other pan. The pair that is heavier contains the heavy canister. Set the lighter pair aside.
- Finally, place one canister from the heavier pair on the other pan. This last measurement will identify the heavy canister in a total of three uses of the balance.

## Using the balance twice:

Place three canisters on each pan of the balance. This should identify one of the three groups of three as the heaviest group. Select this group and place one canister on each pan and one of the three groups of three as the heaviest group. Select this group and place one canister on each pan and one on the bench. If the two pans balance, the heavy canister is on the bench.

# The Puzzle of Nine

## Student Sheet

Imagine that you have been given nine objects, identical in size shape and appearance. Eight of the objects have the same mass, while the ninth one has a slightly greater mass than the others. The difference in mass is so small that it can be detected only by using a balance. You have a double pan balance, but **you may use it only TWO times.**

Your challenge is to design and describe a procedure to identify the object with greater mass.

---

Planning Notes:

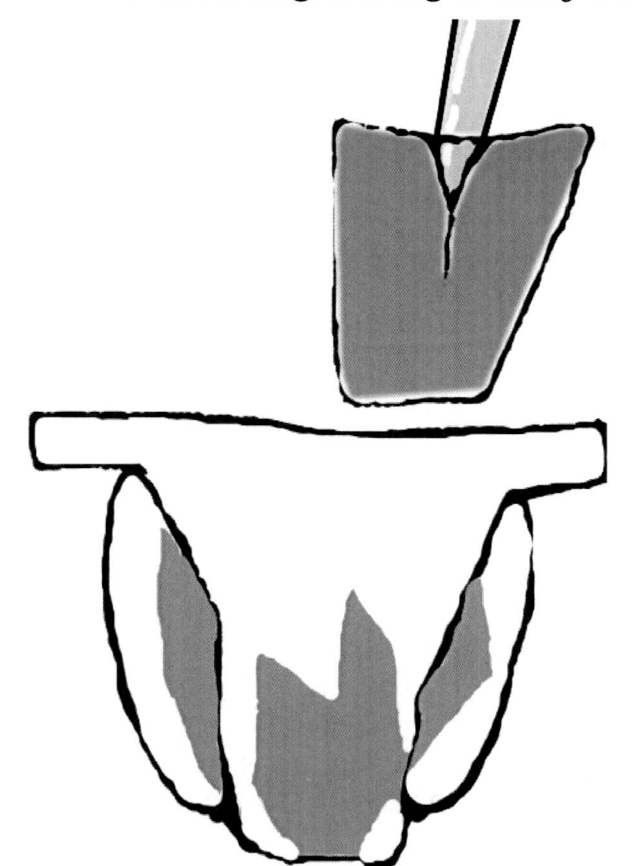

# DIGGING INTO DIAPERS

## Demonstrating the Power of Polymers within Diapers

Time Required: 30-40 minutes

### Standards Alignment Benchmarks

Science:

Level III (Grade 6-8):

| | |
|---|---|
| Standard 8: | Understands the structure and properties of matter |
| Benchmark 8: | Knows that substances react in characteristic ways with other substances to form new substances (compounds) with different characteristic properties |
| Standard 11: | Understands the nature of scientific knowledge |
| Benchmark 2: | Understands the nature of scientific explanations (e.g., use of logically consistent arguments; emphasis on evidence; use of scientific principles, models, and theories; acceptance or displacement of explanations based on new scientific evidence) |
| Standard 12: | Understands the nature of scientific inquiry |
| Benchmark 5: | Uses appropriate tools (including computer hardware and software) and techniques to gather, analyze, and interpret scientific data |
| Benchmark 6: | Establishes relationships based on evidence and logical argument (e.g., provides causes for effects) |

Level IV (Grade 9-12):

| | |
|---|---|
| Standard 8: | Understands the structure and properties of matter |
| Benchmark 5: | Knows that the physical properties of a compound are determined by its molecular structure (e.g., constituent atoms, distances and angles between them) and the interactions among these molecules |
| Standard 12: | Understands the nature of scientific inquiry |

## How This Learning Experience Meets the National Science Education Standards:

As a result of activities in grades 5-8, all students should develop:

## Content Standard A: Science as Inquiry

### Abilities Necessary to Do Scientific Inquiry

- Use appropriate tools and techniques to gather, analyze, and interpret data.
- Develop descriptions, explanations, predictions, and models using evidence.
- Think critically and logically to make the relationships between evidence and explanations.
- Recognize and analyze alternative explanations and predictions.

### Understanding Scientific Inquiry

- Technology used to gather data enhances accuracy and allows scientists to analyze and quantify results of investigations.
- Scientific explanations emphasize evidence, have logically consistent arguments, and use scientific principles, models, and theories.
- Scientific investigations sometimes result in new ideas and phenomena for study, generate new methods or procedures for an investigation, or develop new technologies to improve the collection of data.

## Content Standard B: Physical Science

### Properties and Changes of Properties in Matter

- A substance has characteristic properties, such as density, a boiling point and solubility, all of which are independent of the amount of the sample; a mixture of substances can often be separated into the original substances using one or more of the characteristic properties.
- Substances react chemically in characteristic ways with other substances to form new substances (compounds) with different characteristic properties; in chemical reactions, the total mass is conserved.
- There are more than 100 known elements that combine in a multitude of ways to produce compounds, which account for the living and nonliving substances that we encounter.

## Content Standard G: History and Nature of Science

### Science as a Human Endeavor

- Women and men of various social and ethnic backgrounds engage in activities of science, engineering, and related fields; some scientists work in teams, and some work alone, but all communicate extensively with others.
- Science requires different abilities, depending on such factors as the field and study and the type of inquiry.

### Nature of Science

- Scientists formulate and test their explanations of nature using observation, experiments, and theoretical and mathematical models.
- It is part of the scientific inquiry to evaluate the results of scientific investigations, experiments, observations, theoretical models, and the explanations proposed by other scientists.

## Background:

Farmers and parents of infants have at least one thing in common. They both need to manage moisture. Farmers need to keep moisture in the soil to promote germination of seeds and root development in the plants they grow. Parents need to keep their infant's cradle clean and dry. Both farmers and parents get help from the same source—a white powder called a superabsorbent polymer. This polymer is a white powder that can absorb water many times its volume, forming a stiff gel. The powder is incorporated in soil and in diapers to retain moisture.

A polymer is a material with molecules (smallest particles) that form a long chain of repeating units. In the super absorbent polymer, each of these units has a portion that holds an electrical charge. The electrical charges on the polymer attract water molecules and bind them to the polymer. Each charge binds several water molecules, and each molecule of the polymer has thousands of these charges. Therefore, a small volume of polymer can bind a large volume of water. When salt is added to the mixture of water and polymer, the polymer releases the water. This occurs because salt also contains electrical charges. The charges in salt also attract water molecules, and water molecules bind to the salt instead of the polymer. Furthermore, the charges in salt are attracted to the charged parts of the polymer and displace the molecules of water from the polymer.

Super absorbent polymers have a wide variety of uses, both in industry and in consumer products. Besides being used in baby diapers, its ability to bind water to itself has led to its use in filtering fuels for automobiles and jets. Here it removes small amounts of water that otherwise might freeze and cause blockages in fuel lines. When the super absorbent polymer is distributed in sandy soil, it supports agriculture by improving the soil's ability to retain moisture.

## Teacher Notes:

You may want to use guiding questions when working with the students and their observations of the powder. The following are examples of questions that would be appropriate:

- What color is the powder?
- Is the powder fine or coarse?
- What does the powder resemble?

The students can also use time measurement when making their observations. Things they can focus on include the following:

- How long does it take for the liquid powder mixture to thicken so that it no longer pours?
- What does the thickened mixture look like?

As the students add salt to the mixture, the following questions can be used:

- After adding salt, what happens as you stir?
- Does the mixture change so that is will pour again?
- What happens if you add more salt and stir again?

When removing the salt from the super absorbent material, it is possible to eventually have the original material from the diaper remain after the evaporation. This material can be used again to absorb more water. This is an extension that the students can test while doing long-term observations.

## Safety Notes:

Wear safety goggles while working with the super absorbent powder. Be careful not to inhale dust or fibers from the filling when extracting the powder from the diapers. Wear disposable gloves if needed.

## Getting Started:

1. Determine whether each student, pair, or group of students will engage in the experience; it can be used as a demonstration only for the whole group, but is easily done with any configuration of students.
2. Purchase enough diapers, based on the objectives for the students; each demonstration will require super absorbent disposable diapers (Ultra Pampers is an example of this type of diaper).
3. Gather the remaining materials for the experience.
4. Based on the level of the students, the powder may need to be extracted from the diapers in advance; the extraction is done as follows:
   - Cut open the super absorbent diapers.
   - Place the cotton-like filling into a five-gallon plastic bag; the filling will feel gritty because of the super absorbent powder.
   - Seal the bag; by manipulating through the bag, shred the diaper filling until it is in small pieces.
   - Rub the filling against itself then shake the bag to loosen the powder from the filling; watch for the powder to settle to the bottom of the bag.
   - Open the bag and feel the filling; if it still feels gritty, close the bag and rub the filling again.
   - When the filling no longer feels gritty, gently shake the bag to settle the powder to the bottom.
   - Open the bag and remove the filling, taking care not to inhale dust or fibers from the filling
   - If needed, the powder on the bottom of the bag can be separated from most of the remaining pieces of filling by passing it through the kitchen sieve.
5. Copy student sheets.

## Materials Needed to Extract Super Absorbent Powder:

- 3 super absorbent disposable diapers per group of students
- Very large zip lock bags (must be able to seal; no smaller than one gallon in size)
- Cutting implement
- Kitchen sieve
- Storage containers for extracted powder

## Materials Needed Per Pair or Group of Students:

- Super absorbent powder from 3 disposable diapers
- Water*
- Measuring spoons
- Paper towels
- Safety goggles
- Disposable gloves (optional)
- 2 glasses
- Table salt*
- Spoon
- Student sheets
- Kitchen sieve
- Dust Mask (critical for asthmatics)

* Amounts will vary based on student needs and teacher objectives; plan amounts required based on number of students involved as well as depth of exploration and investigation.

## Procedure:

1. Assemble students into cooperative pairs or groups.
2. Distribute materials; if students are to extract their own super absorbent powder, have them wear dusk masks and follow the instructions above.
3. Have students investigate the super absorbent powder as follows:
   - Place the powder in an empty, dry glass.
   - Observe the powder; record observations on the student sheet.
   - Fill the second glass with water.
   - Pour the water from the second glass into the glass with the powder.
   - Pour the mixture back and forth from one glass to another.
   - Record your observations on the student sheet.
   - Sprinkle a teaspoon of table salt on top of the thickened mixture, stirring with a spoon.
   - Record what happens on your student sheet.
   - Pour the mixture of powder, water, and salt into a sieve lined with paper towels.
   - Slowly run water through the sieve.
   - Observe what happens as the water washes away.
   - Predict what will happen if the rinsing is stopped, the excess water is pressed out, and the contents are placed in an open dish for several days.

# DIGGING INTO DIAPERS
## Student Sheet

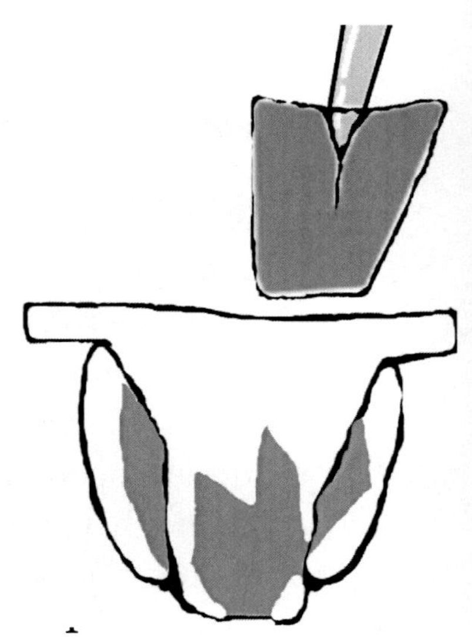

Complete your investigation as follows:

- Place the powder in an empty, dry glass.
- Observe the powder and record your observations.
- Fill the second glass with water.
- Pour the water from the second glass into the glass with
- Pour the mixture back and forth from one glass to another.
- Record your observations.
- Sprinkle a teaspoon of table salt on top of the thickened mixture, stirring with a spoon.
- Record what happens.
- Pour the mixture of powder, water, and salt into a sieve lined with paper towels.
- Slowly run water through the sieve.
- Observe what happens as the water washes away.
- Predict what will happen if the rinsing is stopped, the excess water is pressed out, and the contents are placed in an open dish for several days.

RECORD OBSERVATIONS ON THE BACK OF THIS SHEET

# EXTRACTING SUPER ABSORBENT POWDER
## Student Sheet

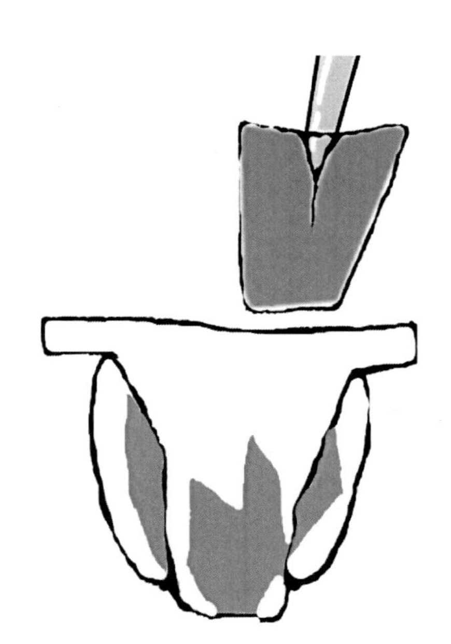

- Cut open the super absorbent diapers.
- Place the cotton-like filling into a five-gallon plastic bag; the filling will feel gritty because of the super absorbent powder.
- Seal the bag; by manipulating through the bag, shred the diaper filling until it is in small pieces.

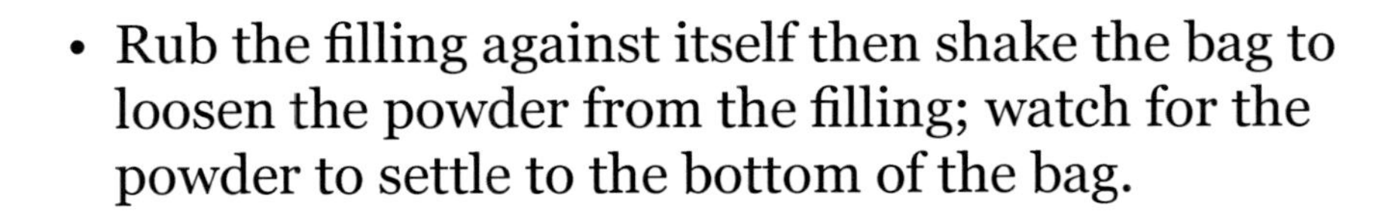

- Rub the filling against itself then shake the bag to loosen the powder from the filling; watch for the powder to settle to the bottom of the bag.
- Open the bag and feel the filling; if it still feels gritty, close the bag and rub the filling again.
- When the filling no longer feels gritty, gently shake the bag to settle the powder to the bottom.
- Open the bag and remove the filling, taking care not to inhale dust or fibers from the filling.
- If needed, the powder on the bottom of the bag can be separated from most of the remaining pieces of filling by passing it through the kitchen sieve.

RECORD OBSERVATIONS ON THE BACK OF THIS SHEET

# THE BOUNTIFUL BAG

Time Required: 60-75 minutes including experimentation and discussion

## How this Learning Experience Meets the National Science Education Standards:

As a result of activities in grades 5-8, all students should develop:

### Content Standard A: Science as Inquiry

#### Abilities Necessary to Do Scientific Inquiry

- Identify questions that can be answered through scientific investigations.
- Design and conduct a scientific investigation.
- Use appropriate tools and techniques to gather, analyze, and interpret data.
- Develop descriptions, explanations, predictions, and models using evidence.
- Think critically and logically to make the relationships between evidence and explanations.
- Recognize and analyze alternative explanations and predictions.
- Communicate scientific procedures and explanations.
- Use of mathematics in all aspects of scientific inquiry.

#### Understanding Scientific Inquiry

- Different kinds of questions suggest different kinds of scientific investigations.
- Mathematics is important in all aspects of scientific inquiry.
- Technology used to gather data enhances accuracy and allows scientists to analyze and quantify results of investigations.
- Scientific explanations emphasize evidence, have logically consistent arguments, and use scientific principles, models, and theories.
- Scientific investigations sometimes result in new ideas and phenomena for study, generate new methods or procedures for an investigation, or develop new technologies to improve the collection of data.

## Content Standard B: Physical Science

### Properties and Changes of Properties in Matter

- A substance has characteristic properties, such as density, a boiling point and solubility, all of which are independent of the amount of the sample; a mixture of substances can often be separated into the original substances using one or more of the characteristic properties.
- Substances react chemically in characteristic ways with other substances to form new substances (compounds) with different characteristic properties; in chemical reactions, the total mass is conserved.

## Content Standard G: History and Nature of Science

### Science as a Human Endeavor

- Women and men of various social and ethnic backgrounds engage in activities of science, engineering, and related fields; some scientists work in teams, and some work alone, but all communicate extensively with others.
- Science requires different abilities, depending on such factors as the field and study and the type of inquiry.

### Nature of Science

- Scientists formulate and test their explanations of nature using observation, experiments, and theoretical and mathematical models.
- It is part of the scientific inquiry to evaluate the results of scientific investigations, experiments, observations, theoretical models, and the explanations proposed by other scientists.

# Introduction:

The Bountiful Bag allows students to practice observational skills and note the difference between observation and inference. It also provides the student with the opportunity to design and perform simple inquiry investigations in order to answer a question posed by the instructor. By combining specific amounts of sodium bicarbonate, calcium chloride, and phenol red in a closed zip lock bag system, the learning experience focuses the student on the following concepts: physical properties of substances, evidence of chemical change, exothermic versus endothermic reactions, reactants, and products.

# Background Information:

## The Reaction

The need for you or your students to understand the precise chemical events that take place in this reaction will depend upon the level of student understanding prior to the learning experience and the objectives you set forth when planning. Scientific observation of the reaction itself is the essence of this learning experience. However, should there be a need for more concept development for higher level students, the following information is provided to enhance the learning experience. ***In any and all cases, DO NOT present this information to the students prior to the learning experience!***

- Calcium chloride, sodium bicarbonate (baking soda) and a liquid combine to produce a carbon dioxide gas
- Calcium chloride and water produce heat
- Phenol red is an acid-base indicator; it changes color in the presence of acids and bases. Baking soda is a base, so it maintains the bright color of the phenol red at the beginning of the reaction. As acidic products form, the phenol red changes color from bright pink to orange, and then yellow. Carbon dioxide gas is one of the acidic products formed. If the gas is allowed to escape, the liquids may turn slightly orange or pink again.

The results of the individual experiments should be as follows:

- $NaHCO_3$ + phenol red $\rightarrow$ cold
- $CaCl_2$ + phenol red $\rightarrow$ hot
- $CaCl_2 + NaHCO_3 + H_2O \rightarrow$ hot
- $NaHCO_3 + H_2O \rightarrow$ cold
- $CaCl_2 + H_2O \rightarrow$ hot

Some of the products in this reaction include:

- Sodium chloride (NaCl): table salt
- Calcium carbonate ($CaCO_3$): the main component of chalk
- Carbon dioxide ($CO_2$): one of the gases we exhale

The events that take place in the zip lock bag are part of a dynamic and fairly complex reaction. Many individual events occur to form a series of intermediate products. High school teachers may want to challenge their students to determine the steps of the reaction and balance the equation.

The following notation represents what takes place:

$CaCl_2 + 2NaHCO_3 \rightarrow Ca\ (HCO_3)2 + 2NaCl$

$Ca(HCO_3)_2 \rightarrow CaCO_3 + CO_2 + H_2o$

(according to a disproportionate process by which a substance, in this case $Ca(HCO_3)2$ undergoes oxidation and reduction at the same time)

## Teacher Notes:

Facilitating inquiry-based learning can be demanding in that it can require tremendous amounts of materials as well as enormous blocks of time. "The Bountiful Bag" enables the student to engage in an inquiry-based learning experience while keeping supplies needed, as well as time required, to a minimum. Multiple concepts are reinforced or introduced, as dictated by the objectives stated, and the students are fully engaged in hands-on/minds-on learning that facilitates connections in science.

Further questions about "The Bountiful Bag" experience should be developed and tested as deemed appropriate by the instructor. Students can also develop independent investigations that allow the students to test different variable as they search for the answer to a question or solution for a problem.

## Safety Notes:

The following safety issues should be discussed with students prior to initiating the learning experience:

- Wear safety goggles at all times during the learning experience.
- Do not touch any chemical.
- Remember to waft when determining if a chemical has an odor. (A demonstration of the wafting technique may be required—hold the open container way from the face and with your free hand sweep the air above the container toward the nose to detect an odor.)
- Zip lock bags may explode if amounts of chemicals are increased.
- All experiments must be approved by the teacher before being conducted.
- If you come into contact with a chemical, flush the area with water. (Calcium chloride can be a skin irritant.)
- Dispose of all zip lock bags without opening them. (The solution is acidic and can irritate eyes and skin.)

•Wash hands following the learning experience.

## Getting Started:

1. Prior to this learning experience, students should be able to successfully do the following:
    - Make observations and draw inferences.
    - Independently form hypotheses.
    - Engage in experimental design to conduct investigations in search of solutions to problems.
    - Differentiate between physical and chemical properties.

- Identify characteristics of a chemical change.
- If the teacher expects the student to draw inferences based on pH, basic knowledge of acids, bases, and indicators must be attained.
- If the student is to successfully balance the accompanying chemical equation, understanding of chemical formulas, the Law of Conservation of Matter, and the ability to balance equations is required.

2. Gather materials for the experience.
3. Copy student sheets.
4. Determine most effective means of reporting results for the whole group discussion.

## Materials Needed:

Per cooperative group of four students:

- 6-8 quart sized zip lock bags
- 6-8 medicine measuring cups
- 4 plastic spoons or teaspoon sized measuring spoons
- Toothpicks
- Magnifying lenses
- 100-500 mL bottle phenol red
- A small container of sodium bicarbonate
- A small container of calcium chloride
- Trays for materials
- Safety goggles
- "The Bountiful Bag" Student sheets
- "What Makes It Hot" Student sheets
- Student learning logs or journals

## For Whole Group:

Area for recording student hypotheses and results (dry erase board, overhead transparency, flip chart, etc.)

## Procedure:

1. Prepare trays with appropriate materials for each cooperative group.
2. Review all safety procedures with the students.
3. Have the materials manager for each group obtain a tray of necessary materials and "The Bountiful Bag" student sheet.
4. Instruct the students to begin by carefully making observations of the three reagents on the tray: sodium bicarbonate, calcium chloride, and phenol red; record all observations on the "Reaction in a Baggie" student sheet.
5. Inform the students that they are then to follow the instructions on the student sheet and make further observations, recording them on the sheet provided.
6. After the students have performed the reaction, have them share their results and record them on the board; guide the students through effective questioning to the discovery that a chemical reaction has occurred, and have them cite evidence to support this conclusion (e.g., gas produced, color change, temperature change, bubbling, new substance formed).

7. Pose the following question: "What makes the bag get hot?"; have the students brainstorm possible answers to the question and record them on the board.
8. Give each student a copy of the "What Makes It Hot?" student sheet.
9. Have each group design two different experiments that can be tested in search of an answer to the question posed; inform students that all investigations must be approved before conducted. (Make certain that all possible solutions posed are being tested by the class.)
10. Have students report their group results to the class; record them on the board.
11. Facilitate a discussion that leads the students to the discovery of which chemicals must be mixed in order to produce heat (both calcium chloride and an aqueous liquid are needed to produce the heat; sodium bicarbonate and the liquid will cause the bag to feel cold).
12. Introduce the terms "exothermic" and "endothermic" as part of the concept development to explain the differences in temperature. (An exothermic reaction produces heat and feels hot; an endothermic reaction requires the input of energy in the form of heat and feels cool.)
13. More experienced students can be challenged to balance the chemical equation that correlates to the learning experience and express their understanding of the Law of Conservation of Matter; in addition, connections to pH and acids and bases can be discussed as well as oxidation and reduction reactions.
14. Have students differentiate between exothermic and endothermic reactions as well as physical and chemical change in a learning log or journal entry; understanding of the balanced equation and other connections can be included in the entry as well.

# The Bountiful Bag
## Student Sheet

Record 5 physical properties of each of the following chemicals:

1. Calcium chloride ($CaCl_2$)

2. Baking Soda ($NaHCO_3$)

3. Phenol Red Solution

## Complete the following steps:

1. Put about 8.4 g (5 mL or 1 teaspoon) of baking soda into a zip lock bag.

2. Add about 16.9 g (10 mL or 2 teaspoons) of calcium chloride to the same bag.

3. Mix the two chemicals. Describe what happens.

4. Measure 10 mL of phenol red solution in a medicine measuring cup.

5. Put the container of phenol red solution into the zip lock bag so that the liquid does not spill. Press the air out of the baggie. Seal the zip lock bag.

6. Tip the vial of phenol red solution. Record your observations.

# What Makes It Hot?
# Student Sheet

What is required to make the mixture get hot?

Experiment 1:

We mixed:

We observed:

We concluded:

Experiment 2:

We mixed:

We observed:

We concluded:

Using complete sentences, describe what makes the mixture hot.

# What Makes It Hot?
## Scoring Rubric

| | |
|---|---|
| 0 points | No student sheet turned in |
| 1-3 points | Hypothesis stated (answer to initial question posed) and one experiment partially recorded |
| 4-6 points | Hypothesis stated with one experiment fully recorded |
| 7-9 points | Hypothesis stated, one experiment fully recorded and second experiment partially recorded |
| 10-12 points | Hypothesis stated and both experiments fully recorded |
| 13-14 points | Hypothesis stated, both experiments fully recorded and conclusion stated in an incomplete sentence (answer to second question posed) |
| 15 points | Hypothesis stated, both experiments fully recorded and conclusion stated in a complete sentence |

# THE POP IN YOUR POP!

## Putting the "Fizz" Into Root Beer

Time Required: 30-40 minutes

### How this Learning Experience Meets the National Science Education Standards:

As a result of activities in grades 5-8, all students should develop:

### Content Standard A: Science as Inquiry

#### Abilities Necessary to Do Scientific Inquiry

- Identify questions that can be answered through scientific investigations.
- Use appropriate tools and techniques to gather, analyze and interpret data.
- Develop descriptions, explanations, predictions and models using evidence.
- Think critically and logically to make the relationships between evidence and explanations.
- Recognize and analyze alternative explanations and predictions.
- Communicate scientific procedures and explanations.
- Use mathematics in all aspects of scientific inquiry.

#### Understanding Scientific Inquiry

- Different kinds of questions suggest different kinds of scientific investigations.
- Mathematics is important in all aspects of scientific inquiry.
- Technology used to gather data enhances accuracy and allows scientists to analyze and quantify results of investigations.
- Scientific explanations emphasize evidence, have logically consistent arguments, and use scientific principles, models, and theories.
- Scientific investigations sometimes result in new ideas and phenomena for study, generate new methods or procedures for an investigation, or develop new technologies to improve the collection of data.

## Content Standard B: Physical Science

### Properties and Changes of Properties in Matter

- A substance has characteristic properties, such as density, a boiling point, and solubility, all of which are independent of the amount of the sample; a mixture of substances can often be separated into the original substances using one or more of the characteristic properties.
- Substances react chemically in characteristic ways with other substances to form new substances (compounds) with different characteristic properties; in chemical reactions, the total mass is conserved.

## Content Standard G: History and Nature of Science

### Science as a Human Endeavor

- Women and men of various social and ethnic backgrounds engage in activities of science, engineering, and related fields; some scientists work in teams, and some work alone, but all communicate extensively with others.
- Science requires different abilities, depending on such factors as the field and study and the type of inquiry.

### Nature of Science

- Scientists formulate and test their explanations of nature using observation, experiments, and theoretical and mathematical models.
- It is part of the scientific inquiry to evaluate the results of scientific investigations, experiments, observations, theoretical models, and the explanations proposed by other scientists.

### History of Science

- Many individuals have contributed to the tradition of science. Studying some of these individuals provides further understanding of scientific inquiry, science as a human endeavor, the nature of science, and the relationships between science and society.
- In historical perspective, science has been practiced by different individuals in different cultures. In looking at the history of many peoples, one finds that scientists and engineers of high achievement are considered to be among the most valued contributors to their culture.
- Tracing the history of science can show how difficult it was for scientific innovators to break through the accepted ideas of their time to reach the conclusions that we currently take for granted.

## Background:

Root beer is now made from the bark of the wild cherry tree and other flavors such as clove, mint and vanilla. If the commercial concentrate is added to water and sugar, something is still missing—the "fizz." Fizz is carbon dioxide forced into a solution by some type of process. The original root beer was made of bark and roots, then fermented so that the yeast slowly produced carbon dioxide in the sealed bottle. However, carbon dioxide can be added to root beer in other ways. As a classroom experience, solid carbon dioxide in the form of dry ice can be immersed into the solution.

## Teacher Notes:

Root beer is fun to make because it serves as a link to the past as well as a chance to produce a beverage that, in modern times, is usually associated with commercial bottling only. This exploration is a good time to capitalize on the cross-disciplinary nature of the activity by researching and discussing history.

## Reaction:

In commercial root beer, the carbon dioxide is forced into the solution under pressure. Gases are more soluble in water if the pressure is increased. Carbon dioxide reacts with water to make carbonic acid:

$$CO_2\,(g) + H_2O\,(l) \rightarrow H_2CO_3\,(aq)$$

Carbon dioxide gas + liquid water yields aqueous carbonic acid

Because the pressure is much lower in the plastic bags than in the commercial process, the root beer produced during the classroom experience will have less dissolved $CO_2$ and thus less "fizz" than commercial root beer. However, students should be easily guided to the right idea about the role $CO_2$ plays in the preparation of most sodas.

## Preparation Notes:

To make a gallon of root beer, mix 12 mL of root beer concentrate (2 tsp), 400 g of sugar (2 cups), and 4 liters of water. Root beer concentrate is available in the spice section of supermarkets. Follow the directions on the bottle.

You will need about one pound of dry ice. It can sometimes be obtained from a grocery store during a "truckload meat sale." The dry ice vaporizes (sublimes) readily so that storage is difficult. If you must store it overnight, buy two pounds, wrap it in brown paper bags, and store it in a polystyrene cooler. Break it with a blunt instrument such as a hammer and HANDLE IT WITH TONGS AND GLOVES. The temperature is -780 C.

If you are going to use cabbage juice as the indicator for acids/bases, prepare it as follows:

- Cut 1⁄4 head of red cabbage into chunks.
- Place in a blender and cover with water.
- Blend thoroughly.
- Strain the juice, disposing of the cabbage pulp.

It is convenient to fill a disposable pipette with cabbage juice in advance for each student team or pair to use when testing the liquids.

## Safety Notes:

The safe handling of dry ice is critical. Students must NOT handle dry ice like they would regular ice. DO NOT PUT DRY ICE IN THE MOUTH! Handle the dry ice with tongs and gloves (if appropriate); DO NOT LET THE DRY ICE COME IN CONTACT WITH THE SKIN!

Be certain that students do not seal the zip lock bags completely after adding the dry ice to the root beer solution; as the dry ice sublimes, or turns to $CO_2$ gas, the pressure will increase and the bag will explode.

Students should wear safety goggles when working; it is recommended that this be done outside of the laboratory area.

## Getting Started:

1. Identify a source for dry ice; each student team (pair) will need two small chunks of dry ice.
2. Review the safe handling of dry ice; be prepared to discuss this with the students prior to beginning the experience.
3. Prepare the root beer solution by mixing the following:
   - 6 mL or 1 tsp. root beer concentrate (available in spice section of supermarkets)
   - 1 cup sugar
   - 2 L water
4. Determine if any concept extensions need to be connected to the making of the root beer.
5. If acids/bases or pH are to be addressed, students may need to review testing acids and bases prior to beginning the root beer experience; if so, do the following:
   - Gather and organize appropriate indicators for testing the pH of solutions.
   - Have vinegar, water and disposable pipettes available.
   - Instruct students to place 100 mL water into one of the zip lock bags and test with the chosen indicator; observe.
   - Have the students add a small amount of vinegar to the water and test with the chosen indicator; observe.
   - Instruct students to place 100 mL of water into another zip lock bag and add a small chunk of dry ice.
   - Test with the chosen indicator and observe.
   - After completing the exploration, compare the observations between the two and determine the relationship between the dry ice and the end result.
6. Gather the remaining materials for the experience.
7. Copy student sheets.

## Materials Needed Per Team/Pair of Students:

- Three or four heavy sandwich-size zip lock bags
- 100 mL graduated cylinder
- 118 mL root beer solution
- 2 small chunks of dry ice
- Clean tongs to handle dry ice
- Disposable pipettes (if appropriate)
- pH indicator or indicator paper (if appropriate)
- Vinegar (if appropriate)*
- Water (if appropriate)*
- Small cups
- Student sheets
- Safety goggles
- Student learning logs

* Amounts will vary based on student needs and teacher objectives; plan amounts required based on number of students involved as well as depth of exploration and investigation.

# Procedure:

1. Assemble students into cooperative groups; identify pairs or teams within the groups.
2. If appropriate, guide students through the acids and indicator exploration as follows:
   - Instruct students to place 100 mL water into one of the zip lock bags and test with the chosen indicator; observe
   - Have the students add a small amount of vinegar to the water and test with the chosen indicator; observe, recording observations in learning logs
   - Instruct students to place 100 mL of water into another zip lock bag and add a small chunk of dry ice
   - Test with the chosen indicator and observe, recording observations in learning logs
   - After completing the exploration, compare the observations between the two and determine the relationship between the dry ice and the end result
3. Have students make root beer using the following procedure:
   - Obtain a clean zip lock bag.
   - Measure out 118 mL of root beer solution; add to the clean bag.
   - Obtain a small chunk of dry ice; add the dry ice to the root beer solution and seal the bag almost completely shut.
   - Allow the reaction to proceed until all of the dry ice is gone.
   - Pour the prepared root beer into small cups and taste.
   - Decide whether the root beer is acidic or not; indicate how you could find out, and record in learning log
4. Facilitate a discussion based on the student responses; develop concepts as needed.

# THE POP IN YOUR POP!
## Student Sheet

Prepare your root beer as follows:

- Obtain a clean zip lock bag.
- Measure out 118 mL of root beer solution; add to the clean bag.
- Obtain a small chunk of dry ice; add the dry ice to the root beer solution and seal the bag almost completely shut.
- Allow the reaction to proceed until all of the dry ice is gone.
- Pour the prepared root beer into small cups and taste.
- Decide whether the root beer is acidic or not; indicate how you could find out, and record in learning log

## Questions To Consider:

1. Does adding carbon dioxide to a solution make it acidic?

2. How do you know?

3. What would happen if the plastic bag were completely sealed with the dry ice in it?

4. What makes soda pop "go flat" if the lid is left off for a time?

5. List some other carbonic acid solutions that are a part of your everyday life.

# THE POP IN YOUR POP!
## Teacher Sheet with Suggested Answers

Prepare your root beer as follows:

- Obtain a clean zip lock bag.
- Measure out 118 mL of root beer solution; add to the clean bag.
- Obtain a small chunk of dry ice; add the dry ice to the root beer solution and seal the bag almost completely shut.
- Allow the reaction to proceed until all of the dry ice is gone.
- Pour the prepared root beer into small cups and taste.
- Decide whether the root beer is acidic or not; indicate how you could find out and record in learning log

## Questions To Consider:

1. Does adding carbon dioxide to a solution make it acidic?
   ***Yes (see reaction noted).***

2. How do you know?
   ***An acid indicator would react the same way (be the same color) in the root beer as it would in another acid solution (like vinegar).***

3. What would happen if the plastic bag were completely sealed with the dry ice in it?
   ***It would explode.***

4. What makes soda pop "go flat" if the lid is left off for a time?
   ***The carbon dioxide is less soluble at the lower pressure and escapes.***

5. List some other carbonic acid solutions that are a part of your everyday life.
   ***Rainwater, blood plasma, mineral springs, pancake batter, etc.***

# Reactionary Rockets!

## Exploring Rocket Launch Through Process Skills and the Scientific Method

Time Required: 45-60 minutes

### How this Learning Experience Meets the National Science Education Standards:

As a result of activities in grades 5-8, all students should develop:

### Content Standard A: Science as Inquiry

Abilities Necessary to Do Scientific Inquiry

- Identify questions that can be answered through scientific investigations.
- Design and conduct a scientific investigation.
- Use appropriate tools and techniques to gather, analyze and interpret data.
- Develop descriptions, explanations, predictions and models using evidence.
- Think critically and logically to make the relationships between evidence and explanations.
- Recognize and analyze alternative explanations and predictions.
- Communicate scientific procedures and explanations.
- Use mathematics in all aspects of scientific inquiry.

Understanding Scientific Inquiry

- Different kinds of questions suggest different kinds of scientific investigations.
- Mathematics is important in all aspects of scientific inquiry.
- Technology used to gather data enhances accuracy and allows scientists to analyze and quantify results of investigations.
- Scientific explanations emphasize evidence, have logically consistent arguments, and use scientific principles, models, and theories.
- Scientific investigations sometimes result in new ideas and phenomena for study, generate new methods or procedures for an investigation, or develop new technologies to improve the collection of data.

## Content Standard B: Physical Science

### Properties and Changes of Properties in Matter

- Substances react chemically in characteristic ways with other substances to form new substances (compounds) with different characteristic properties.

### Motions and Forces

- The motion of an object can be described by its position, direction of motion, and speed; that motion can be measured and represented on a graph.
- An object that is not being subjected to a force will continue to move at a constant speed and in a straight line.
- If more than one force acts on an object along a straight line, then the forces will reinforce or cancel one another, depending on their direction and magnitude; unbalanced forces will cause changes in the speed or direction of an object's motion.

## Content Standard E: Science and Technology

### Understanding Science and Technology:

- Scientific inquiry and technological design have similarities and differences.
- Technological designs have constraints.
- Technological solutions have intended benefits and unintended consequences.

## Content Standard G: History and Nature of Science

### Science as a Human Endeavor

- Women and men of various social and ethnic backgrounds engage in activities of science, engineering, and related fields; some scientists work in teams, and some work alone, but all communicate extensively with others.
- Science requires different abilities, depending on such factors as the field and study and the type of inquiry.

### Nature of Science

- Scientists formulate and test their explanations of nature using observation, experiments, and theoretical and mathematical models.
- It is part of the scientific inquiry to evaluate the results of scientific investigations, experiments, observations, theoretical models, and the explanations proposed by other scientists.

**Did you Know?**

Mae C. Jemison, a mission specialist on the Endeavor, launched September 12, 1992, was the first African-American woman in space.

## Overview:

This learning experience incorporates the scientific method, chemistry and physical science. Students must determine the best "recipe" for launching Alka Seltzer® rockets the greatest distance. Exploring the possibilities gives the students opportunities to practice observing, communicating, measuring, inferring, predicting, controlling variables, and collecting and analyzing data.

## Background:

The independent variable is the one that is changed on purpose—this is the variable that the experimenter actually manipulates throughout the investigation. It causes the change that will be measured. That change is called the dependent variable—this is the variable that responds and is measured. The constants are kept the same throughout the entire experiment.

The solid antacid tablet has two ingredients that do not react with each other when they are solid: citric acid and sodium bicarbonate. However, both of these ingredients dissolve in water and when they do, a chemical reaction takes place between them. Carbon dioxide gas is produced. During the learning experience, the carbon dioxide is trapped inside the film canister. As the carbon dioxide produced increases, the pressure inside the canister builds until the force pops the lid and launches the "rocket."

## Teacher Notes:

The challenge for students is to find the combination of factors that will cause the rocket to travel the greatest distance. Variables to test include: amount of water in the rocket, temperature of the water, size of the tablet (whole, half, etc.?), tablet in one piece or crushed, and angle of launch.

It is recommended that teachers demonstrate how to correctly launch the film canister rockets by doing the following:

- Place an alka seltzer tablet and water in a film canister.
- Snap on the lid.
- Load the rocket, cap first, into the PVC elbow.
- Point the rocket away from students.
- Wait for the blast-off.

Student data can be graphed and shared as well. Following the completion of the rocket launch, student graphs can be used in a whole-group carousel. Each group then analyzes the data shown in each graph and forms conclusions prior to the sharing of results and the discussion that follows. In addition, students can be encouraged to construct their own data table in their learning log rather than use the prepared sheet.

## Safety Notes:

Students should be instructed to wear their safety goggles. There should be a designated area for launching the rockets. The rockets must never be aimed at another student or group. Solutions may be washed down the drain.

## Getting Started:

1. Prepare the PVC rocket launchers using the instructions provided.
2. Purchase fresh Alka Seltzer® tablets if needed; tablets that have been subjected to the air for long periods of time should not be used.
3. Gather the remaining materials needed for the group investigations.
4. Copy student instruction sheets and data sheets.
5. Review the follow up questions to be used when facilitating the discussion.
6. Determine how the investigation will be best implemented in the classroom—will everyone do the same test first then each group conduct a separate test, will each group begin with separate tests, etc.
7. Identify an appropriate space for the rocket launch to take place.
8. If further content other than the scientific method is to be developed, review concepts and organize the most effective connection to the investigation.

## Materials Needed Per Group:

- Effervescing tablets (any brand will work)
- Clear film canister
- Graduated cylinder
- Water
- Hot pot (or other source for warm water)
- Thermometer
- Meter stick (or metric measuring tape)
- Rocket launcher (1" PVC elbow)
- Student instruction sheets
- Student data sheets
- Safety goggles
- Learning logs
- Graph boards or paper (optional)

## Procedure:

1. Assemble students into cooperative groups.
2. Demonstrate the rocket launcher without providing any information; have the students record observations and explanations in their learning logs.
3. Facilitate a discussion of what occurred using student input and reflections; guide students toward listing what factors could impact the "rocket" performance.
4. Post a list on the board of the combination of factors that could cause a rocket to travel the greatest distance.
5. Assign each group a specific factor or variable to test.
6. Using the student sheets as a guide, discuss the procedure to follow; encourage students to make notes in their learning logs as needed.
7. Have materials managers gather the supplies for the investigation; point out the appropriate place for the rocket launch to take place.
8. Assign a time limit for the investigation to be completed; if time allows, encourage students to test another variable in an effort to increase the distance their rocket will travel.
9. At the completion of the investigation, have each group share its results; facilitate a discussion and develop content as needed.

# Reactionary Rockets
# Student Sheet

## Materials:

- Alka-Seltzer® tablets (any brand will work)
- Clear film canister
- Graduated cylinder
- Water
- Hot pot (or other warm water source)
- Thermometer
- 1” PVC elbow
- Meter stick

## Instructions:

Your group’s task is to determine the combination of factors that will launch your rocket the greatest distance. Decide which variable your group will test and the steps you will take. Repeat the test three times to get an average. If time permits, you may then change another variable to try to increase your distance even more.

## Describe Your Experiment:

(Include your data chart and measurements.)

## Reactionary Rocket Recipe
## Student Sheet

| "I" Changed Variable (independent) | Responding Variable (dependent) | | |
|---|---|---|---|
| | | | |
| | Trial 1 | Trial 2 | Trial 3 |
| | | | |
| | | | |
| | | | |
| | | | |

# Reactionary Rockets
# Follow Up Questions

When the groups have completed their investigations, have each one report about what they did and their results. As a class, discuss:

1. Did the amount of water make a difference? If so, was more or less better?

2. Did the temperature of the water make a difference? If so, was colder or warmer better?

3. Did the amount of Alka-Seltzer make a difference? If so, was more or less better?

4. Did it make a difference whether the tablet was whole or crushed? If so, which was better?

5. Did launch angle affect distance traveled? If so, what appeared to be the best angle (estimate)?

6. Based on the class data, what should be the ultimate combination of factors to send the rocket the greatest distance?

7. Would we need to conduct more experiments to verify your choice?

8. Could our procedures be improved? How?

# STRETCHING THE TRUTH

## Chewing Up a Polymer Investigation

Time Required: 45-60 minutes

### How this Learning Experience Meets the National Science Education Standards:

As a result of activities in grades 5-8, all students should develop:

### Content Standard A: Science as Inquiry

#### Abilities Necessary to Do Scientific Inquiry

- Identify questions that can be answered through scientific investigations.
- Design and conduct a scientific investigation.
- Use appropriate tools and techniques to gather, analyze and interpret data.
- Develop descriptions, explanations, predictions and models using evidence.
- Think critically and logically to make the relationships between evidence and explanations.
- Recognize and analyze alternative explanations and predictions.
- Communicate scientific procedures and explanations.
- Use mathematics in all aspects of scientific inquiry.

#### Understanding Scientific Inquiry

- Different kinds of questions suggest different kinds of scientific investigations.
- Mathematics is important in all aspects of scientific inquiry.
- Technology used to gather data enhances accuracy and allows scientists to analyze and quantify results of investigations.
- Scientific explanations emphasize evidence, have logically consistent arguments, and use scientific principles, models, and theories.
- Scientific investigations sometimes result in new ideas and phenomena for study, generate new methods or procedures for an investigation, or develop new technologies to improve the collection of data.

## Content Standard B: Physical Science

### Properties and Changes of Properties in Matter

- A substance has characteristic properties, such as density, a boiling point, and solubility, all of which are independent of the amount of the sample.

## Content Standard G: History and Nature of Science

### Science as a Human Endeavor

- Women and men of various social and ethnic backgrounds engage in activities of science, engineering, and related fields; some scientists work in teams, and some work alone, but all communicate extensively with others.
- Science requires different abilities, depending on such factors as the field and study and the type of inquiry.

### Nature of Science

- Scientists formulate and test their explanations of nature using observation, experiments, and theoretical and mathematical models.
- It is part of the scientific inquiry to evaluate the results of scientific investigations, experiments, observations, theoretical models, and the explanations proposed by other scientists.

# Background:

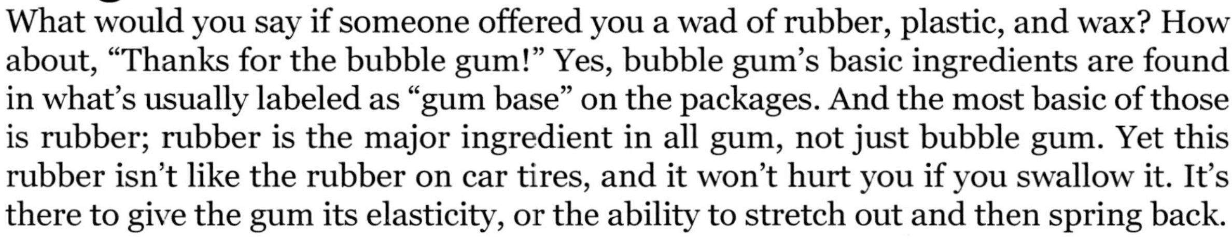

What would you say if someone offered you a wad of rubber, plastic, and wax? How about, "Thanks for the bubble gum!" Yes, bubble gum's basic ingredients are found in what's usually labeled as "gum base" on the packages. And the most basic of those is rubber; rubber is the major ingredient in all gum, not just bubble gum. Yet this rubber isn't like the rubber on car tires, and it won't hurt you if you swallow it. It's there to give the gum its elasticity, or the ability to stretch out and then spring back.

But if gum base were just rubber, then "brewing a chew" would be like chewing on erasers. So gum chemists add other ingredients, notably plastics and waxes. The plastic combines with the rubber to make gum base easier to chew; waxes keep the gum hard in the package. However, in your mouth your body heat "melts" the waxes, softening the gum.

Now, if someone just asked you to chew on a wad of rubber, plastic, and wax, you would probably refuse. It would have to taste really good in order to win you over. That's why gum chemists add flavorings and loads of sugar to their products. A few minutes after you start chewing, these substances dissolve in your saliva, giving you a high-power burst of flavor.

So why aren't all gums the same? Some blow great bubbles; other break apart as soon as you attempt to blow the bubble. The reason is different amounts of rubber. Bubble gum has a higher concentration of rubber. And it turns out that rubber is where the real secret of "blowability" lies.

If you could take a molecular-level look at the rubber in gum base, you'd see that it is made of giant molecules called polymers. A polymer is made of many smaller, identical molecules joined like the links in a chain. In gum rubber, the polymers are arranged in a loose "tangle" of overlapping chains.

The structure may not be neat, but it makes the rubber stretchable. Scientists have a name for the stretchable (elastic) polymers. They call them elastomers.

How does the elastomer in gum base stretch? When you blow a bubble, the force of your breath straightens the polymer chains into long, orderly bundles and the bubble expands. But that force strains the chemical bonds in the chains. If one of the bonds breaks, POP! The bubble "springs a leak," the air rushes out, and the straightened chains recoil into tangles. In regular gum, with a lower concentration of stretchy elastomer, the bonds break sooner. Sometimes this happens even before you can "pucker up." Thus, no bubbles.

The size of the bubble you can blow actually depends on the non-elastic ingredients that gum chemists add. Plastic is a good example; more plastic and less elastomer mean less stretch. And, you could certainly make a bubble gum that would inflate to the size of a basketball. But when the bubble broke, the gum would wrap around your head! So gum makers must limit bubbles to the size of softballs.

## Bubble Gum Tidbits:

- The energy Americans spend every day chewing bubble gum could light a city of ten million people.
- Over the years, Americans have chomped on enough bubble gum to make a stick nearly 113 million miles long; that's long enough to reach the moon and back more than 200 times.
- Scientists say bubble gum chewing can help you work off muscle tension.
- If teenaged taste testers say "no go" to a new product, chemists go back to the drawing board.
- Bubble gum makes a great Earth-friendly insecticide; pests nibble the stuff and their jaws stick together for good.
- In San Luis Obispo, California, people have added wad after wad to a gum mural in town, forming a gum "pop art."

## Teacher Notes:

It is imperative to discuss proper experimental design with students before engaging them in the gum investigation. Students must be made aware of the need to test only one variable at a time. Based on the level of the students, you may want to allow them to design their own experimental procedure. If the students are not ready for developing their own procedure, an active discussion with good guiding questions can lead them toward the procedure you have chosen for them to complete. In addition, the students can work through a card sort where they are given the steps to the procedure and must place them in logical order before ever beginning the investigation. A template for the card sort is included in this learning experience.

Be sure to emphasize the parts of the experimental design and clarify the difference in the independent and dependent variables. Graphing skills should be reinforced as well.

If measuring implements are scarce, students can use dental floss to identify the length of each trial. Instead of pulling the gum along side a measuring implement, the gum can be pulled along side a roll of dental floss. When the gum breaks, the dental floss can be cut to match the length of the stretched gum. The piece of dental floss can then be measured and the length recorded. This will also prevent gum from getting stuck to the measuring tapes or sticks. If using the dental floss, emphasize to the students the need to keep the pieces from being mixed up when measuring and recording the data.

## Safety Notes:

Allergies to gum or gum products must be addressed first. Provide disposable gloves for students to use when handling the chewed gum. If gloves are not available, have students hold the gum with a small piece of wax paper. Also be sensitive to the needs of diabetic students or those who have related disorders.

## Getting Started:

1. Check for gum allergies within the whole group before beginning; if permission slips need to be distributed and signed by parents or guardians, do so.
2. Determine whether students will work in pairs or groups.
3. Decide how many different brands of gum each pair/group of students will test.
4. Purchase a variety of brands and types of gum.

5. Gather the remaining materials needed for the investigation.
6. If using the experimental procedure card sort, prepare sets of cards for each group or pair of students.
7. Be prepared to discuss experimental variables with the students as well as appropriate methods that can be used to conduct the test.
8. Identify an appropriate space for the tests to be conducted. Area should be free of obstacles or items that could be damaged by the gum.

## Materials Needed Per Group of Students:

- Several brands of bubble gum
- Balance or scale
- Timer
- Measuring tape or meter stick
- Knife or implement for cutting gum
- Calculator
- Graph paper (optional)
- Student sheets
- Disposable gloves (optional)
- Experimental procedure card sets (optional)

## Procedure:

1. Assemble students into cooperative groups; assign pairs within the groups if appropriate.
2. Discuss testing different brands of bubble gum with the students; guide them toward a procedure that is similar to the following (use the card sort sets if appropriate):
   - Measure out four grams of gum; be sure to test only ONE brand of gum at a time.
   - Chew the four gram portion for three minutes.
   - Remove the wad of chewed gum from the mouth and roll it into a ball.
   - Hold the ball in one hand next to the zero end of the measuring implement.
   - Have another student grasp both the gum and measuring implement, then back away slowly until the gum breaks.
   - Record how far the gum stretched before breaking in the table provided.
   - Repeat the procedure three times for each brand of gum.
   - Calculate the average stretch distance for each brand of gum tested.
   - Prepare an appropriate graph that represents your results.
3. Discuss the experimental results with the students; as a group, make predictions for the following questions:
   - Will the gum that stretched the farthest make the biggest bubbles?
   - Which stretches more, regular or sugar-free gum?
   - Does the length of chewing time affect bubble size?
   - Does the temperature of the gum affect bubble size?
4. Engage students in brainstorming other questions about the gum that could be answered after appropriate experimentation; facilitate further investigations if time allows.

# STRETCHING THE TRUTH
## Student Sheet

Get to the truth about gum and its "stretchability" by doing the following:

- Measure out four grams of gum; be sure to test only ONE brand of gum at a time.
- Chew the four gram portion for three minutes.
- Remove the wad of chewed gum from the mouth and roll it into a ball.
- Hold the ball in one hand next to the zero end of the measuring implement.
- Have another student grasp both the gum and measuring implement, then back away slowly until the gum breaks.
- Record in the table provided how far the gum stretched before breaking.
- Repeat the procedure three times for each brand of gum.
- Calculate the average stretch distance for each brand of gum tested.
- Prepare an appropriate graph that represents your results.

# STRETCHING THE TRUTH DATA

| The Effect of Gum Brand on Stretch Distance | | | | |
|---|---|---|---|---|
| Gum Brand | Maximum Stretch Distance (measured in cms) | | | Average Length (sum of three trials divided by three) |
| | Trail #1 | Trial #2 | Trial #3 | |
| Ex: Dirty Bubble | 108 cm | 120 cm | 113 cm | 114 cm |
| | | | | |
| | | | | |
| | | | | |
| | | | | |
| | | | | |
| | | | | |

Graph Your Results below (or use graph paper if available):

## STRETCHING THE TRUTH
### Experimental Procedure Card Sort

| | |
|---|---|
| Measure out four grams of gum. | Be sure to test only ONE brand of gum at a time. |
| Chew the four gram portion for three minutes. | Remove the wad of chewed gum from the mouth. |
| Roll the chewed gum into a ball. | Hold the ball in one hand next to the zero end of the measuring implement. |
| Have another student grasp the other end of the gum and measuring implement. | Back away slowly until the gum breaks. |
| Record in the table provided how far the gum stretched before breaking. | Repeat the procedure three times for each brand of gum. |
| Calculate the average stretch distance for each brand of gum tested. | Prepare an appropriate graph that represents your results. |

# Chemical Engineering Word Search Puzzle

K Q E O W T F G N I R U T C A F U N A M E T Z E V K
H R D J J K I Q M I M O Y Y Q X S A I U S J A K H J
L Y V E F R K H F A I W G F M S E I R A L A S K A G
E E O N O Y F X T C N H R W M T P C R E T R V V A U
L N B F M A A E K I Y A E B F I Z X F M O Z R C K Y
E R I I E I R V Q P R N N J D F F T V G Y Q D Y M Q
C O O V G I K I S N G D E E N V I R O N M E N T A L
T T C N A R X B E T E N B N U K C T C B D L A T R A
R T H L L A J R V T M Q S E O V R D A Q M P V E R M
O A E G B M L A I U Y O L Q L P P I Q J O X P N J Y
N V M U N H G T T S R O T C O D C E T O N F E N H T
I U I Z H A D I A H D V I T Q H B P I P E K F I Q P
C M C G A M C O E Q Q U S V E R U V J P Y N M S B Y
S I A Z M X C N R F P H A R M A C E U T I C A L D D
C X L G A W G S C U J F A W K L K E E W E D W L N I

Chemical engineering experts that reduce
ution and clean our drinking water.

Chemical engineers work in the
_________ industry.
Chemical engineers are ___________.

Chemical engineers work in _________
ence.
A chemical engineer can become a
cialized ___________.
Chemical engineers make ____________
es.
American Institute of Chemical Engineers
Chemical engineers can also be _________.

2. ___________ companies employ a large number of chemical engineers to research, develop or design their product lines.
4. Another industry that employs chemcial engineers.
6. Chemical engineers who focus on biomedical applications.
8. The "sweet spot" (sports equipment) reduces ____________.
10. Chemical engineering offers one of the highest starting ____________.
12. The Department of __________ has become a prominent employer of chemical engineers.
14. National Engineers Week
16. Chemical engineers make good _________.

# Chemical Engineering Word Search Puzzle
# Teacher Sheet

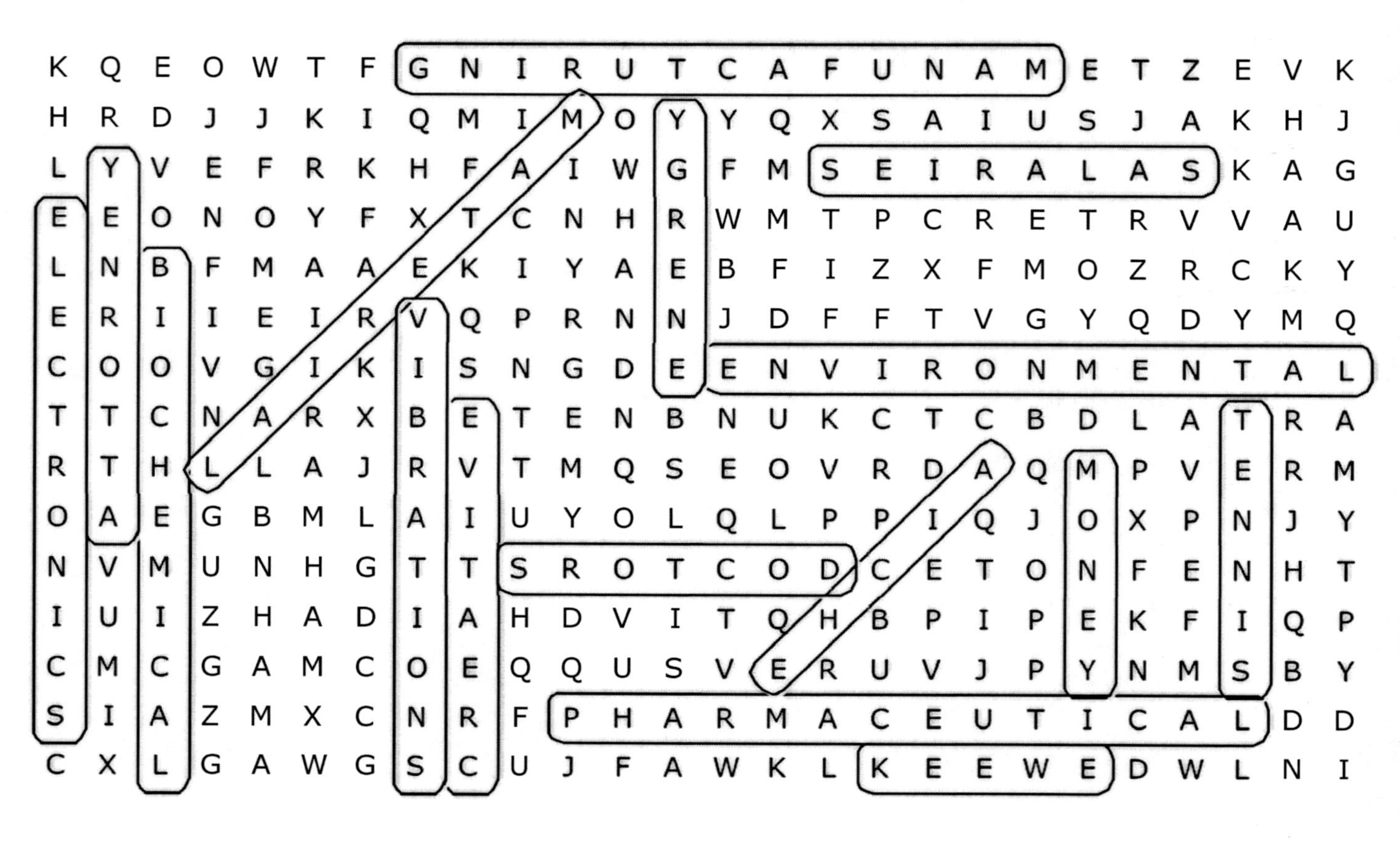


Chemical engineering experts that reduce
›llution and clean our drinking water.
:nvironmental]

Chemical engineers work in the
__________ industry. [electronics]
Chemical engineers are __________.
:reative]
Chemical engineers work in __________
:ience. [material]
A chemical engineer can become a
›ecialized __________. [attorney]
l. Chemical engineers make __________
ıoes. [tennis]

3. American Institute of Chemical Engineers
\IChE]
5. Chemical engineers can also be __________.
loctors]

2. __________ companies employ a large number of chemical engineers to research, develop or design their product lines. [pharmaceutical]
4. Another industry that employs chemcial engineers. [manufacturing]
6. Chemical engineers who focus on biomedical applications. [biochemical]
8. The "sweet spot" (sports equipment) reduces __________. [vibrations]
10. Chemical engineering offers one of the highest starting __________. [salaries]
12. The Department of __________ has become a prominent employer of chemical engineers. [energy]
14. National Engineers Week [eweek]

16. Chemical engineers make good __________. [money]

# Chemical Engineering Crossword Puzzle

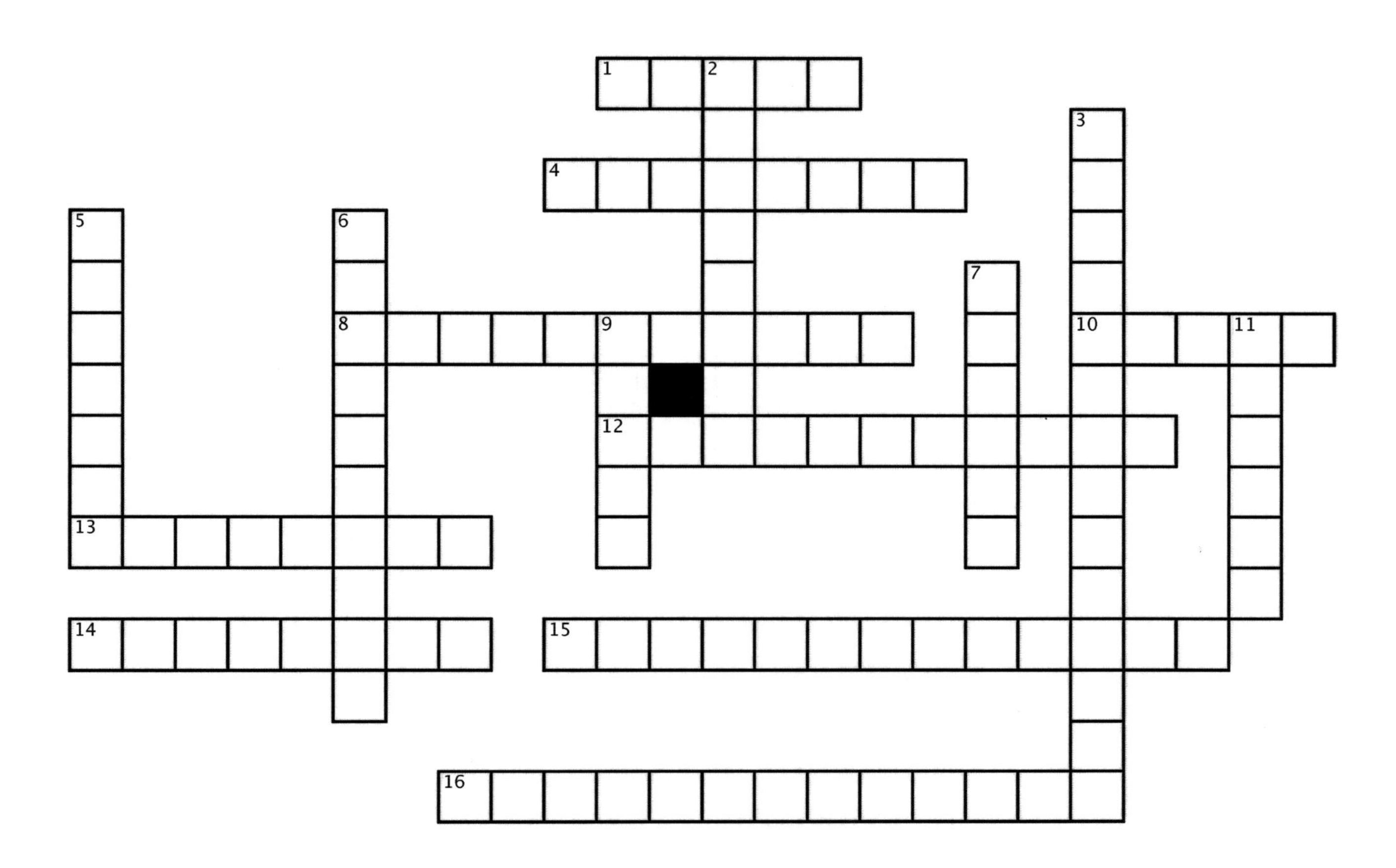


pharmaceutical energy creative salaries tennis vibrations biochemical manufacturing AIChE material doctors attorney money environmental eweek electronics

## Across

1. American Institute of Chemical Engineers
4. Chemical engineers work in __________ science.
8. Chemical engineers who focus on biomedical applications.
10. Chemical engineers make good _________.
12. Chemical engineers work in the ____________ industry.
13. Chemical engineering offers one of the highest starting ____________.
14. A chemical engineer can become a specialized ___________.
15. Another industry that employs chemcial engineers.
16. Chemical engineering experts that reduce pollution and clean our drinking water.

## Down

2. Chemical engineers are ___________.
3. ___________ companies employ a large number of chemical engineers to research, develop or design their product lines.
5. Chemical engineers can also be _________.
6. The "sweet spot" (sports equipment) reduces _____________.
7. Chemical engineers make ____________ shoes.
9. National Engineers Week
11. The Department of __________ has become a prominent employer of chemical engineers.

# Chemical Engineering Crossword Puzzle
## Teacher Sheet

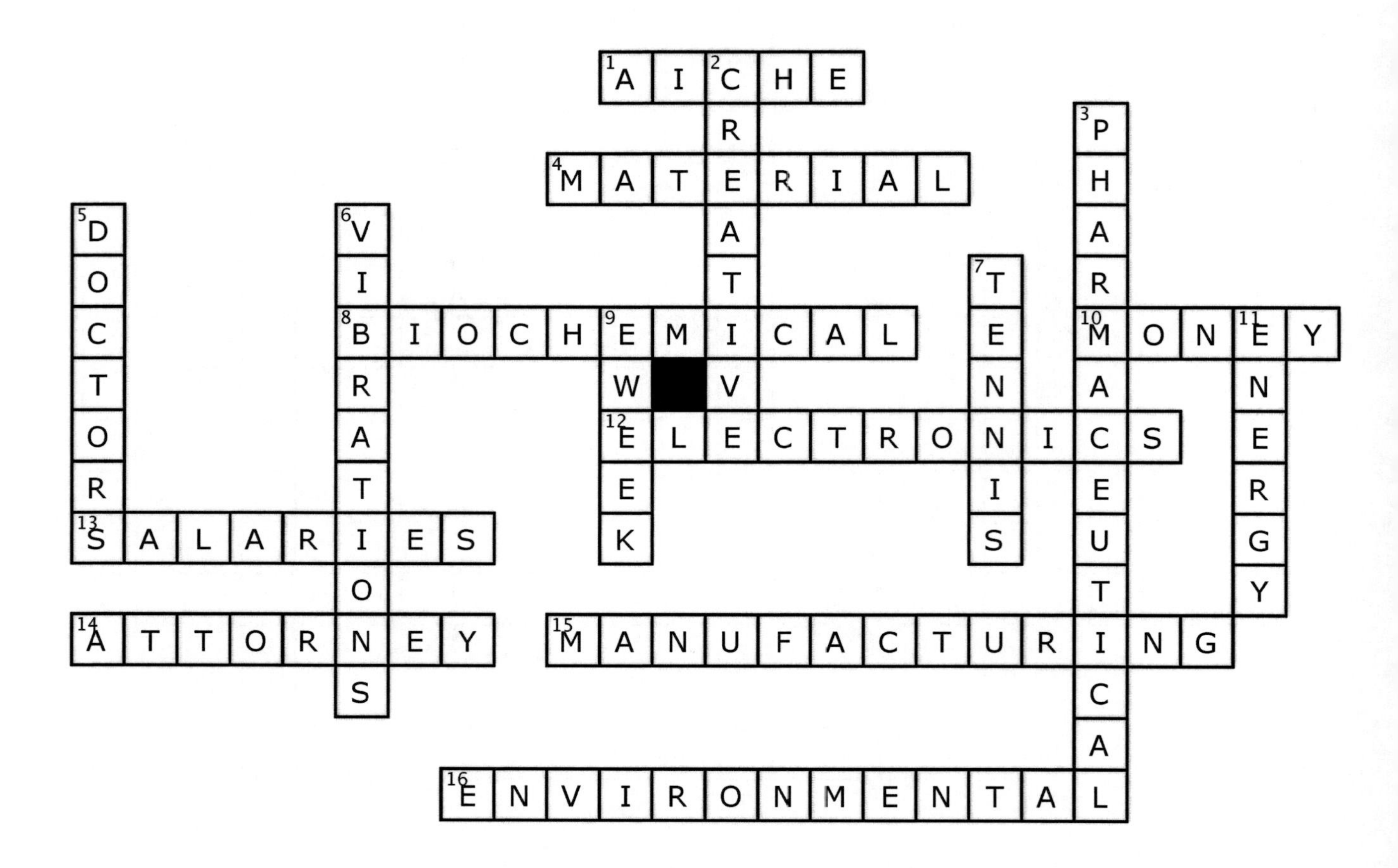


pharmaceutical energy creative salaries tennis vibrations biochemical manufacturing AIChE material doctors attorney money environmental eweek electronics

### Across

1. American Institute of Chemical Engineers [AIChE]
4. Chemical engineers work in __________ science. [material]
8. Chemical engineers who focus on biomedical applications. [biochemical]
10. Chemical engineers make good _________. [money]
12. Chemical engineers work in the ____________ industry. [electronics]
13. Chemical engineering offers one of the highest starting ____________. [salaries]
14. A chemical engineer can become a specialized ___________. [attorney]
15. Another industry that employs chemcial engineers. [manufacturing]

### Down

2. Chemical engineers are ___________. [creative]
3. ___________ companies employ a large number of chemical engineers to research, develop or design their product lines. [pharmaceutical]
5. Chemical engineers can also be _________. [doctors]
6. The "sweet spot" (sports equipment) reduces ____________. [vibrations]
7. Chemical engineers make ____________ shoes. [tennis]
9. National Engineers Week [eweek]
11. The Department of __________ has becom a prominent employer of chemical engineers. [energy]

# Mechanical Engineering

Median income: $79,742

Mechanical engineers are the wheels of the world. This type of engineering is one of the broadest and most diverse disciplines. Every object you used today was the handiwork of a mechanical engineer. Their creations make an impact on all of us and not many people can perform their jobs without them. Mechanical engineers design, develop, and manufacture every kind of vehicle, power system, machine and tool: jet engines, steam engines, power plants, underwater structures, tractors for food production, hydraulic systems, transportation systems, medical devices, sports equipment, smart materials, materials and structures for space travel, measurement devices and more. Any type of machine that produces, transmits, or uses power is most likely the product of a mechanical engineer. They can work in testing, quality assurance, manufacturing, production, sales or product maintenance. There is hardly any aspect of life that has not been influenced by a mechanical engineer.

Job opportunities for mechanical engineers will grow over the next seven years, according to an assessment of the profession released by the American Society of Mechanical Engineers (ASME). Today, an estimated 16 percent more mechanical engineers are employed in the U.S. than in 1996. This percent increase translates to a gain of 35,000 jobs, largely in the services sector of the U.S. economy, over the last 10 years.

These trends include the creation of new materials featuring remarkable attributes of strength and lightness, the miniaturization of medical instruments and other tools, flexible and programmable manufacturing systems allowing rapid switching from one product to another, and the ever increasing role of personal computers in engineering design and analysis. As well, an increasing number of mechanical engineers will be needed to develop the fuel cell and other energy technologies and to manage environmental waste and hazards in compliance with stringent state and federal regulations.

According to the ASME, the job opportunities for mechanical engineers are outstanding. Nationwide, demand for all types of engineers remains well above the average for all other professions. And the mechanical engineers are in even greater demand than most other types of engineers. The U.S. Department of Labor expects that America will need as many as 87,000 new mechanical engineers by 2009.

High demand also means greater opportunities to do interesting work. Because there is so much mechanical engineering work to go around, may young mechanical engineers usually have their choice of good jobs at the beginning of their careers.

High demand for mechanical engineers also means the flexibility to do the work you like to do, and live where you want to. Since machines and mechanical systems are almost everywhere, mechanical engineers can usually find employment where they want to live.

Internationally, expanding populations and growing economies will spur job creation in environmental pollution control and the electric power, telecommunications, and airline industries. Other work opportunities will be found in the high-tech area of vehicle guidance systems to relieve automotive traffic congestion in many parts of the world.

Since mechanical engineering is such a broad discipline, students should select a school with an area of emphasis that matches their own interests. For example, students interested in automotive engineering that wish to broaden their choice of schools should select a college that teaches mechanical engineering with an emphasis in automotive engineering, such as the University of Illinois, Michigan or Tennessee. If your primary interest is to make cars or any vehicle go faster, choose a school with an emphasis in combustion, materials, fluid mechanics or thermodynamics.

Stephen Katsaros, a mechanical engineering student from Purdue with an entrepreneurial spirit, has developed a tool for sharpening alpine skis, a bicycle overhead storage system (BOSS), and, using his creativity and mechanical advantage, is at work on several other inventions. Katsaros revealed a surprising lesson: "College is the best time for an individual to start a business because of all the available resources: the professors, the students to bounce ideas off of, the computers. . . and at Purdue we had a full machine shop."

The ASME Web site is packed with excellent information for students who are interested in mechanical engineering. The Society produces The Mechanical Advantage, a student magazine that tells about competitions, scholarships, industry news, grant programs, and job market forecasts. See what other mechanical engineers are doing every day, and become a member of this progressive and supportive society. Check them out at www.asme.org.

## Julie Dawson
## Mechanical Engineer

From the time I was young I have been inventing things. For example, when I needed a pencil holder, I just made one out of whatever was available. This usually included paper, tape, and cardboard. In fact, I kept a stash of cardboard under my bed so I would always have a supply close by. Some of my inventions could have been purchased at the store, but I knew I would get exactly what I wanted and it would be more fun if I just made it myself. That is how I knew I wanted to be an engineer.

My favorite classes in middle school and high school were science and math where I could learn how the world worked. I loved learning about biology and physics because they talked about things that were happening around me. I enjoyed math because I could use it to solve puzzles.

These classes helped me when I began studying mechanical engineering in college. Of all the disciplines of engineering, I picked mechanical because it allowed me to design. I have a very artsy side to my personality and every time I did a project I was creating something new and unique. I have always been amazed that I can start with an idea and end up with a real physical part.

Currently, I work for a company designing orthopedic implants. For example, if you break your arm and need surgery, I design the plates and screws to put the bone back together. When I start to design a new product, I first learn everything I can about the body. I learn the shape of the bone, what nerves are close by, and other ways of treating the broken bone. This helps me learn different surgical techniques and the benefits and drawbacks of each of them. Additionally, sometimes I go into the operating room with surgeons to learn what is difficult about fixing the bone and what they like and don't like about other products. I use what I have learned to design a product that will be good for both the patient and the doctor. My main goal is to make a product that will help patients get better faster while being easy for the surgeons to use.

I like the way mechanical engineering allows me to be creative and to design new products. It is a very rewarding career because I use these skills to improve the quality of life of other people.

# Mechanical Engineers and some of the things they design:

### Animatronics (i.e. Lucas films and Pixar):

Mechanical engineers may design sets or animatronics/hydraulic systems.

### Athletic shoes:

Mechanical engineers may design systems for manufacturing, motion analysis or impact testing. They may also be involved in building and testing prototypes.

### Baseball and football equipment:
They may design pitching machines, new bat, ball or helmet testing methods, systems for manufacturing, motion analysis or impact testing. They may also be involved in building and testing prototypes.

### Bicycles:
Mechanical engineers may design the frames, derailleur, hubs, forks, handlebars, brakes, spindles, sprockets and everything in between.

### Bowling:
Mechanical engineers may design pin resetters, ball returns, lanes, scoring systems, ball testing systems and bowling alleys.

### Fishing rods and reels/equipment:
They may design rods and reels, reel seats, grips, guide sets or the manufacturing processes.

### Golf equipment:
Mechanical engineers may design golf carts, ball or club manufacturing machines. They may find new ways to produce clubs, and may design new clubs.

### Helmets:
Mechanical engineers may design systems for manufacturing, motion analysis or impact testing. They may also be involved in building and testing prototypes.

### Pools and equipment for swimmers:
Mechanical engineers may design swimming treadmills or other mechanical systems that increase swimming performance.

### Roller Coasters:
Mechanical engineers may design the wheel systems, tracks and drive systems for pulling the roller coaster up the first hill.

### Skateboards:
Mechanical engineers may design the boards, trucks, wheels, bearings and everything in between.

### Skates:
Mechanical engineers may design the frames, wheels, bearings and everything in between.

### Skis and snowboards:
Mechanical engineers may design snow makers. They may design systems for manufacturing, motion analysis or impact testing. They may also be involved in building and testing prototypes.

### Tennis:
Mechanical engineers may design tennis racquets or the machines that produce them. They may also be involved in manufacturing processes of racquets, ball launchers or line-call systems (determines whether a ball is hit in or out of play).

# THE MARK OF SUCCESS

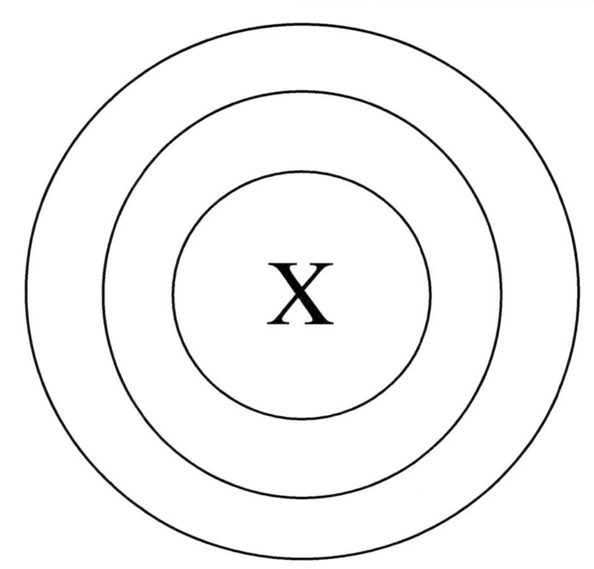


## An Exploration into the Concepts of Accuracy and Precision

Time Required: 45-60 minutes

## How this Learning Experience Meets the National Science Education Standards:

As a result of activities in grades 5-8, all students should develop:

### Content Standard A: Science as Inquiry

#### Abilities Necessary to Do Scientific Inquiry

- Identify questions that can be answered through scientific investigations.
- Design and conduct a scientific investigation.
- Use appropriate tools and techniques to gather, analyze and interpret data.
- Develop descriptions, explanations, predictions and models using evidence.
- Think critically and logically to make the relationships between evidence and explanations.
- Recognize and analyze alternative explanations and predictions.
- Communicate scientific procedures and explanations.
- Use mathematics in all aspects of scientific inquiry.

#### Understandings About Scientific Inquiry

- Different kinds of questions suggest different kinds of scientific investigations.
- Technology used to gather data enhances accuracy and allows scientists to analyze and quantify results of investigations.
- Scientific explanations emphasize evidence, have logically consistent arguments, and use scientific principles, models, and theories.

### Content Standard B: Physical Science

#### Motions and Forces

- The motion of an object can be described by its position, direction of motion, and speed.
- An object that is not being subjected to a force will continue to move at a constant speed and in a straight line.

## Content Standard E: Science and Technology

### Abilities of Technological Design:

- Design a solution or product.
- Implement a proposed design.
- Evaluate completed technological designs or products.
- Communicate the process of technological design.

### Understandings About Science and Technology:

- Scientific inquiry and technological design have similarities and differences.
- Technological designs have constraints.
- Technological solutions have intended benefits and unintended consequences.

## Content Standard G: History and Nature of Science

### Science as a Human Endeavor

- Women and men of various social and ethnic backgrounds engage in activities of science, engineering, and related fields; some scientists work in teams, and some work alone, but all communicate extensively with others.
- Science requires different abilities, depending on such factors as the field and study and the type of inquiry.

### Nature of Science

- Scientists formulate and test their explanations of nature using observation, experiments, and theoretical and mathematical models.
- It is part of the scientific inquiry to evaluate the results of scientific investigations, experiments, observations, theoretical models, and the explanations proposed by other scientists.

## Background:

Numbers make descriptions more exact. When you measure the mass of an object, or its volume or length, you obtain a number in units. But just how accurate is this number? You have to ask yourself how close to the true value will the measurement be.

No value obtained in experiments will be exact because all measurements are subject to errors—there are instrument errors to consider as well as method errors and human errors. If a measuring device is not scaled or calibrated correctly, instrument errors will occur. There are a number of ways this can occur: a scale may be printed incorrectly on a graduated cylinder; a balance may stick and not read small masses correctly.

Standards are important in measurement; it is imperative to use a method that is agreed on by everyone involved. If measurement is going to be made using a graduated cylinder, for example, then everyone must do it the same way every time if the data is to be reliable.

People inevitably make mistakes and human error can occur by chance or through bias, even when great care is taken. Carelessness will contribute many errors but one can't disregard the fact that many times experimenters will have an idea about what answer they would like to see and bias can be introduced. The ability to recognize one's bias is an important part of all types of research.

However, there are two essential things that must be considered when making and reporting measurement: accuracy and precision. Accuracy refers to the extent to which a measurement approaches the true value and is free from error. If you use a dartboard and bull's-eye analogy, accurate darts would all be fairly close to—but not necessarily in a tight cluster to—the center of the bull's-eye. Precision then relates to how exact or sharply stated the measurement is; you would consider how consistent measurements are with one another rather than with the actual value. On a dartboard, precise marks would all be clustered together but not near the bull's-eye. A measurement made on an extremely fine scale is more precise than a coarse measurement made with a blunt instrument. An analytical balance measures mass with greater precision than a triple beam balance.

## Teacher Notes:

Based on the amount of time you have to develop this learning experience, the following modifications may need to be considered:

- Student groups may not need to carousel through the entire series of procedures; each group may do one or two other procedures before continuing with the learning experience.
- Students may need to be formally assessed on their understanding of accuracy and precision. There are several ways to do this, you may want to engage students in a writing assessment that allows them to discuss the difference between accuracy and precision, or students may simply identify scenarios that illustrate accuracy or precision. Groups may design an independent experiment that demonstrates accuracy and precision, or students could write a detailed procedure that outlines the steps of an experiment.

## Safety Notes:

Because students will be moving throughout the learning environment, the only safety concern would be for the mobility of all participants. Should the group include individuals with special needs, modifications or accommodations will need to be addressed.

## Getting Started:

Students begin the learning experience by considering four different visuals. The visuals can be shared by presenting each student with an individual sheet, using a transparency, or placing large versions of the visuals in front of the whole group. Each of the visuals represents a different arrangement of marks on a "target." Challenge students to compare and contrast the diagrams, labeling each based on their understanding of the illustration. Students should be looking for situations that are either accurate, precise, neither accurate nor precise, precise but not accurate, and so on. Do not introduce the terminology to the students. This should be an experience that allows students to express their understanding and previous experience with the concepts. At the appropriate time, facilitate a discussion with the whole group where students share their ideas about the diagrams. Use the student responses and visuals to then develop the appropriate concepts and introduce the needed terminology. Once the students have been provided a concrete understanding of the terms "accurate" and "precise," proceed with the next phase of the learning experience.

## Materials Needed to Begin:

- Visuals of the target diagrams (large copies or individual student sheets)
- Student learning logs

## Materials Needed Per Group:

- Clay, enough for each cooperative group to have at least a 3" x 1" x 1" piece
- Piece of rope (approximately 1 meter in length)
- Copies of the target
- Meter stick or tape measure
- Learning log
- Paper
- Markers

## Procedure:

1. Assemble the students into cooperative groups.
2. Have the materials manager get the materials needed for the group experience: a piece of clay, a piece of rope, a meter stick (or tape measure), a target, paper, and markers.
3. Instruct each group to prepare five balls of clay, equal in size and shape; different groups may have different sized clay balls, but each group's balls should be uniform.
4. Inform the students that they are to place one of the targets on a flat, hard (non-carpeted) surface such as a floor, table top, or desk.
5. Drop one clay ball as close to the center of the target as possible from a height of no less than one meter (3.3 ft.).
6. Students are to experiment with different ways of dropping the balls onto the target in order to maximize their precision and accuracy.
7. Once the group has come to a consensus on the method that they feel is both accurate and precise, they are to record a detailed procedure of the method in their learning logs; a single copy of the procedure should then be transferred to a piece of paper by the data collector.
8. Using the written procedure, have the students then conduct three separate trials, and record the results of each trial directly on the target using three different colored markers to identify each trial.

9. Have the groups present their procedures and results; facilitate a discussion concerning the accuracy and precision of the results shared. Question students about what might explain any variability between the different groups.
10. Instruct students to organize their work stations, leaving the clay balls, targets, and single written procedure on the table top.
11. Engage student groups in a carousel through the other group stations.
12. Inform students that they are to carefully read the new procedures and conduct three trials using the new procedures and materials; record results and make notes in learning logs concerning ability to adequately perform the assigned task.
13. After completing the carousel, have groups organize their results and notes on ability to follow procedures; engage student groups in a brief presentation that provides an overview of the carousel experience.
14. Facilitate a discussion that focuses on the following:
    - Were there any differences in the accuracy and precision between results from the test they designed and those they engaged in during the carousel?
    - Were there any difficulties in following other written procedures?
    - What were the primary causes of these difficulties?
    - How could difficulty in following someone else's procedure lead to inaccuracy in data collection?

# Diagrams
# Student Sheet

1. Describe the illustration:

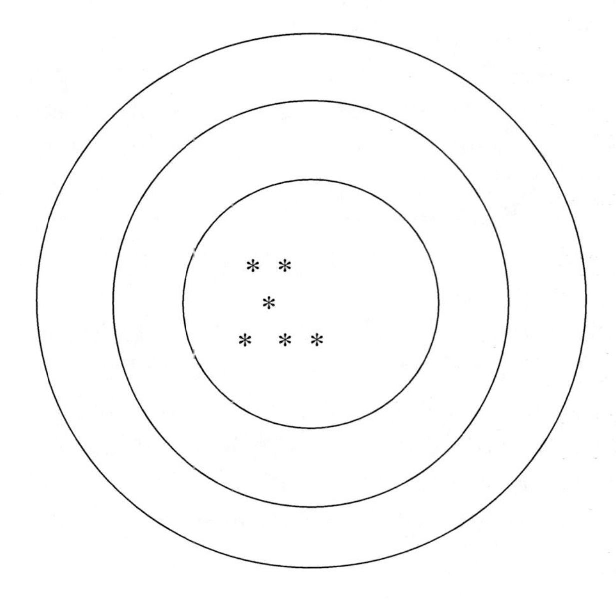

2. Describe the illustration:

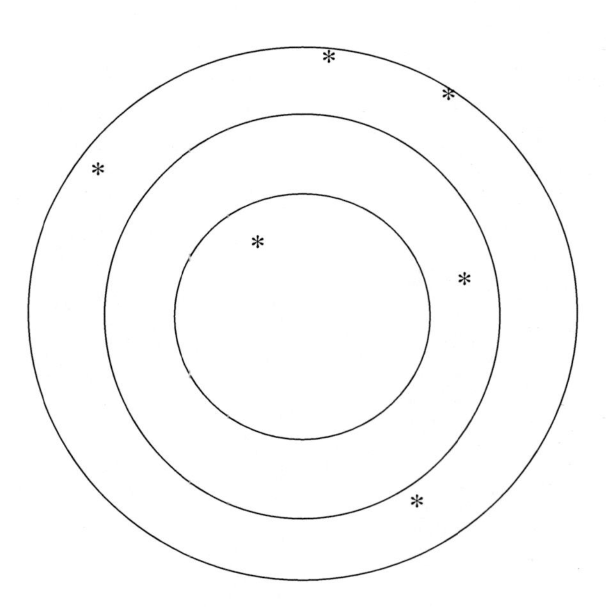

3. Describe the illustration:

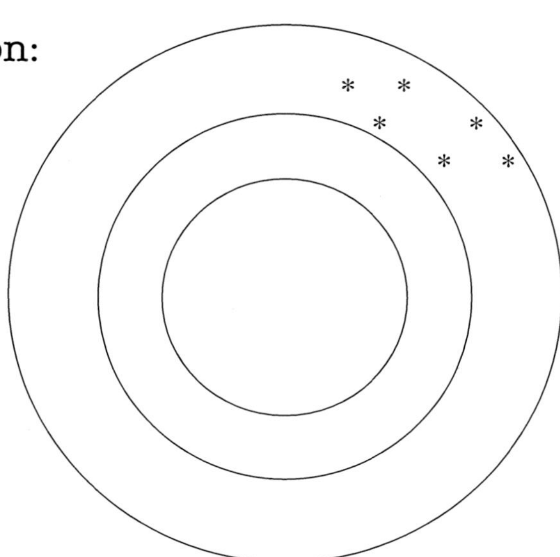

4. Describe the illustration:

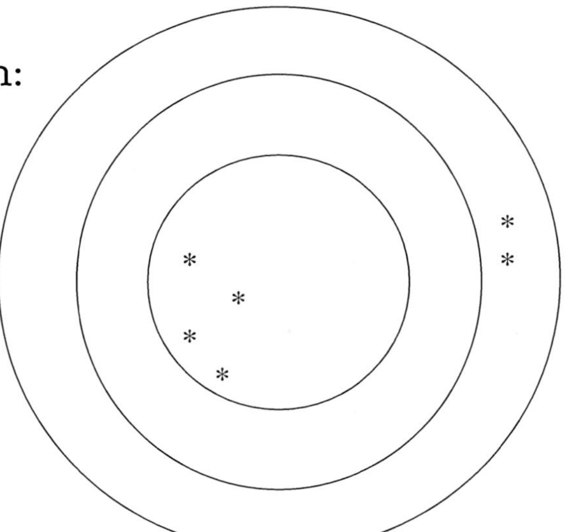

# Diagrams
# Teacher Sheet

1. Describe the illustration:
Good accuracy, good precision

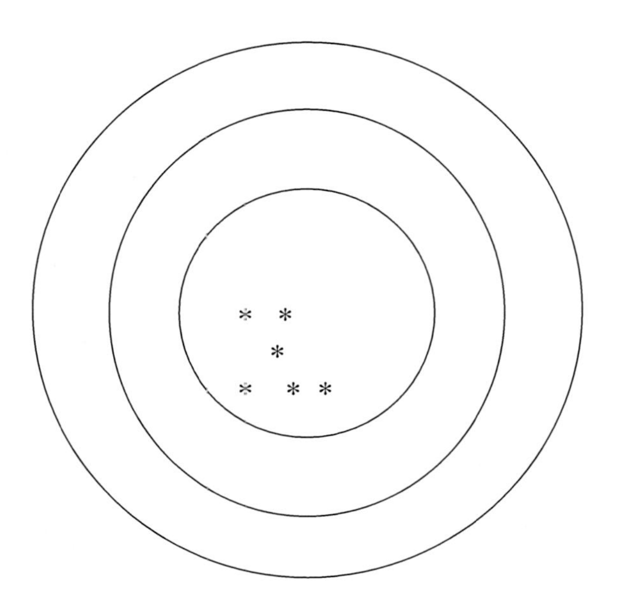

2. Describe the illustration:
One accurate mark; generally poor accuracy overall, poor precision

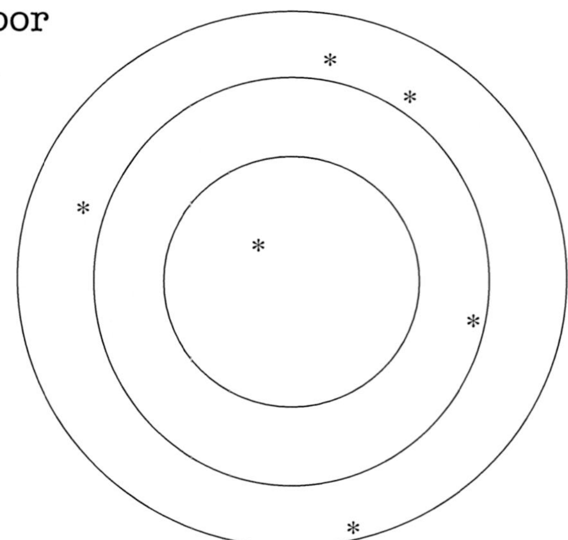

### 3. Describe the illustration:
Good precision, poor accuracy

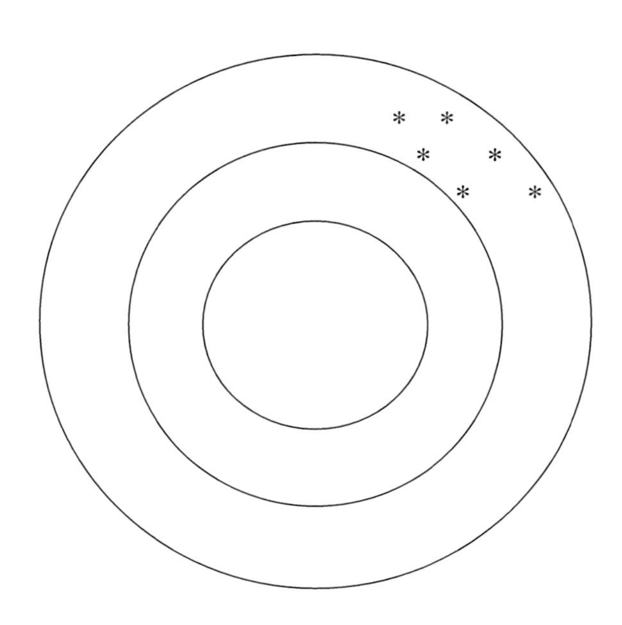

### 4. Describe the illustration:
Some accuracy, some precision; no trend either way— neither accurate nor precise

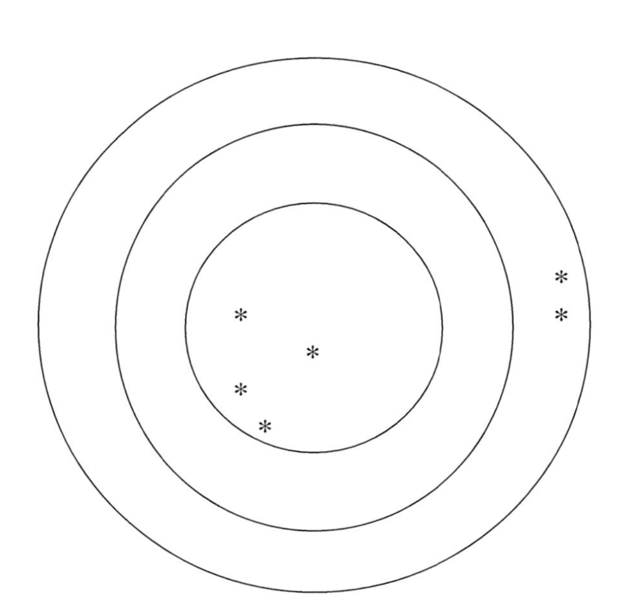

# Group Target

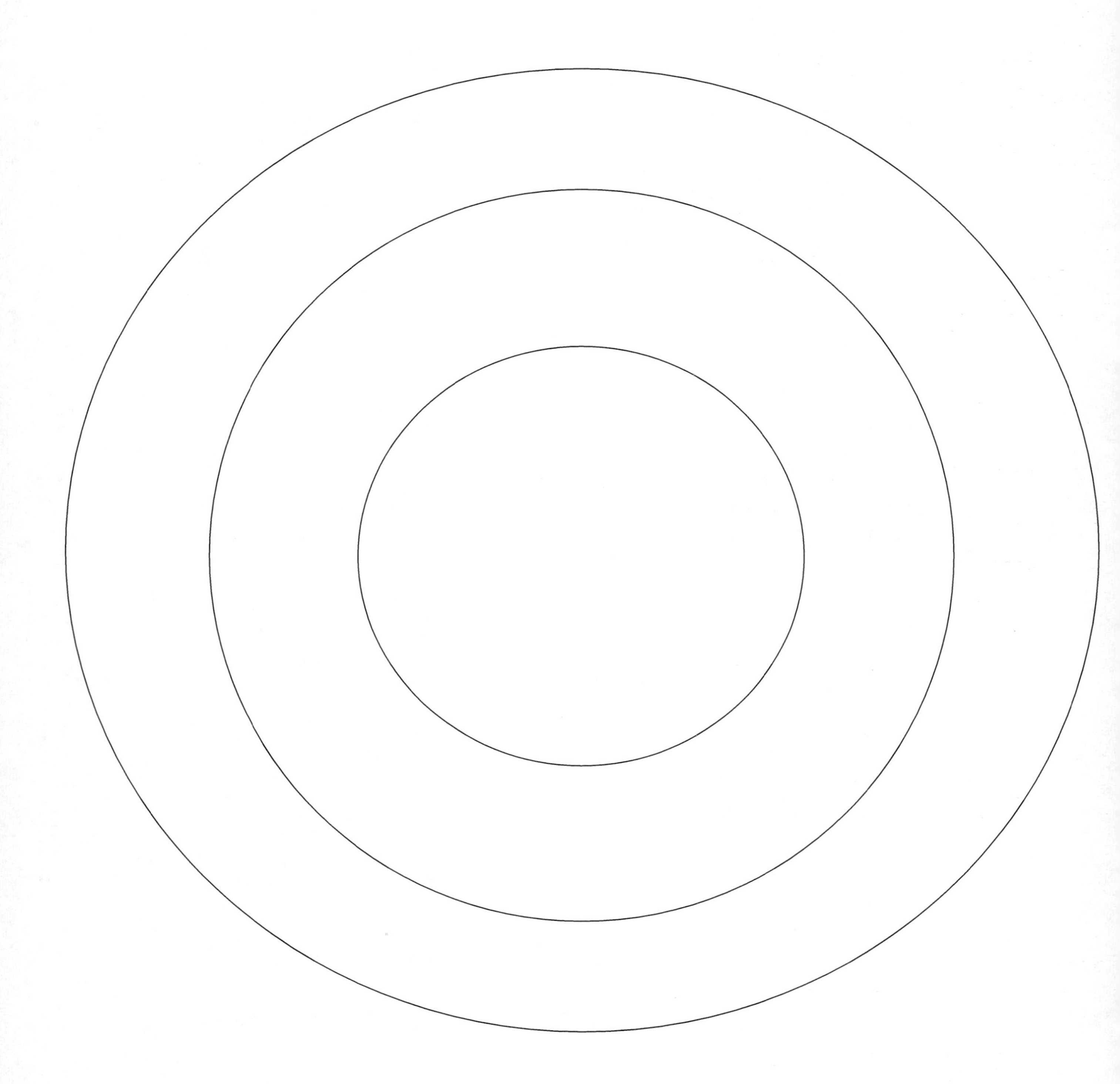

# How Do I Hover?

## An Exploration into How to Lift a Hovercraft

Time Required: 35-40 minutes

### How this Learning Experience Meets the National Science Education Standards:

As a result of activities in grades 5-8, all students should develop:

### Content Standard A: Science as Inquiry

#### Abilities Necessary To Do Scientific Inquiry

- Identify questions that can be answered through scientific investigations.
- Use appropriate tools and techniques to gather, analyze and interpret data.
- Develop descriptions, explanations, predictions and models using evidence.
- Communicate scientific procedures and explanations.
- Use mathematics in all aspects of scientific inquiry.

#### Understanding Scientific Inquiry

- Different kinds of questions suggest different kinds of scientific investigations.
- Mathematics is important in all aspects of scientific inquiry.
- Scientific explanations emphasize evidence, have logically consistent arguments, and use scientific principles, models, and theories.
- Scientific investigations sometimes result in new ideas and phenomena for study, generate new methods or procedures for an investigation, or develop new technologies to improve the collection of data.

### Content Standard B: Physical Science

#### Motions and Forces

- The motion of an object can be described by its position, direction of motion, and speed
- An object that is not being subjected to a force will continue to move at a constant speed and in a straight line.
- If more than one force acts on an object along a straight line, then the forces will reinforce or cancel one another, depending on their direction and magnitude; unbalanced forces will cause changes in the speed or direction of an object's motion.

## Content Standard E: Science and Technology

### Abilities of Technological Design:

- Design a solution or product.
- Implement a proposed design.
- Evaluate completed technological designs or products.
- Communicate the process of technological design.

### Understanding Science and Technology:

- Scientific inquiry and technological design have similarities and differences.
- Technological designs have constraints.
- Technological solutions have intended benefits and unintended consequences.

## Content Standard G: History and Nature of Science

### Science as a Human Endeavor

- Women and men of various social and ethnic backgrounds engage in activities of science, engineering, and related fields; some scientists work in teams, and some work alone, but all communicate extensively with others.
- Science requires different abilities, depending on such factors as the field and study and the type of inquiry.

### Nature of Science

- Scientists formulate and test their explanations of nature using observation, experiments, and theoretical and mathematical models.
- It is part of the scientific inquiry to evaluate the results of scientific investigations, experiments, observations, theoretical models, and the explanations proposed by other scientists.

### History of Science

- In looking at the history of many peoples, one finds that scientists and engineers of high achievement are considered to be among the most valued contributors to their culture.
- Tracing the history of science can show how difficult it was for scientific innovators to break through the accepted ideas of their time to reach the conclusions that we currently take for granted.

## Overview:

This experience enables students to engage in an open inquiry where process skills and higher order thinking are needed in order to complete the challenge. Students work cooperatively to design a small, table-top hovercraft that behaves in the same manner as its real-world counterpart.

## Background:

A hovercraft is a vehicle supported on a cushion of air. These vehicles are designed to travel close to, but above, ground or water. Supported in various ways, some of them have a specially designed wing that will lift them just off the surface over which they travel when they have reached a sufficient horizontal speed (the ground effect). Hovercrafts are usually supported by fans that force air down under the vehicle to create lift; air propellers, water propellers, or water jets usually provide forward propulsion. Air-cushion vehicles can attain higher speeds than can either ships or most land vehicles and use much less power than helicopters of the same weight. Air-cushion suspension has also been applied to other forms of transportation, in particular trains, such as the French Aero train and the British hover train.

## Teacher Notes:

If the students have no previous experience with hovercrafts, it may be helpful to present an appropriate engaging experience. An excellent film montage includes clips from Star Wars, Back to the Future II, and The Matrix Revolutions. This will provide students with a foundation to work from when designing their own hovercraft models

The model can be designed to "hover" by doing the following: After gluing the pop-up top to the CD, pull the top up in the "open" position; fill a balloon with air and without releasing the air, secure the balloon opening over the pop-up top; release the air and watch as the CD "hovers" across the table top, much like an air hockey puck on its table top.

When discussing the homemade hovercrafts, several concepts can be introduced and developed; among them, friction, Newton's laws and aerodynamics. Based on the goals and objectives for the class, you will want to plan a student centered strategy for developing and connecting the concepts related to the learning experience.

There is also a lot of history that can be included in a study of hovercrafts, both history of science and our culture as well. A classroom timeline is an excellent way to incorporate history into your science lesson, adding to it as you cover different topics throughout the year.

Furthermore, NASA has many resources that can be used for the study of hovercraft vehicles and their related concepts, as well as how the concepts affect space travel.

## Safety Notes:

Extreme caution should be used when working with hot glue guns. If students are constructing their own CD apparatus, teachers may still want to handle the gluing process themselves. However, if students will be gluing the caps to the CDs, be sure to demonstrate the appropriate procedure as well as what to do should a burn occur. Have a first aid kit available in case of burns or accidents.

It should also be noted that students tend to get excited when any type of hovercraft begins to move. Therefore, exercise caution in the classroom environment; make sure that students are monitored, aisles and pathways are clear, and students are adequately spaced throughout the work environment.

## Getting Started:

1. Determine whether students will work individually, in pairs, or in cooperative groups.
2. If using an appropriate engaging tool prior to the inquiry, identify it and prepare it as needed.
3. Gather the following materials listed below; the number of each will depend on the configuration of students working together.
4. Based on the level of the students, determine if CD structures need to be prepared in advance or if students are able to construct them individually (or in pairs); if CDs are to be prepared in advance, complete as follows:
   - Place the CD on the table, shiny side up.
   - Position the water bottle top over the center of the CD.
   - Hot glue the water bottle cap to the CD by gluing around the cap to the CD within the ridge.
   - Allow the hot glue to dry for about 10-15 minutes.
5. If students are preparing their own apparatus, plan specific instructions for students to follow while the glue is drying; one idea is to provide them with the challenge of how to make the CD apparatus "hover" or float on air. Record their ideas and sketches in learning logs.
6. Have enough balloons for each CD apparatus to have more than one in case of defects or accidents.
7. Gather assorted materials for the students to choose from when developing their design; suggestions include tape, meter sticks, yarn, plastic wrap, paper clips, craft sticks, foil, clay, pipe cleaners, straws, etc.
8. Make sure students have a flat surface to work on; if appropriate, have a variety of surface types available on which students may explore.
9. Determine how much time students will be given to explore, and test their designs while attempting to make the CD apparatus float above the table top.

## Materials Needed:

- Used or old CDs
- Pop-up tops from dish soap bottles or sports water bottles
- Hot glue gun with glue sticks
- Balloons
- Assorted inquiry supplies (tape, meter sticks, yarn, plastic wrap, paper clips, craft sticks, foil, clay, pipe cleaners, straws, etc.)
- Student learning logs

## Procedure:

1. Assemble students into cooperative groups; indicate whether students are to work in pairs within the group.
2. If appropriate, present the engaging experience.
3. Have materials managers obtain the necessary materials; if students are to begin construction on the apparatus. See steps outlined under "Getting Started."
4. Challenge students to design a way to make the CD apparatus float on air; inform students that only the materials provided can be used and initial ideas, hypotheses, and sketches should be completed in learning logs.

5. Present the materials available but do not allow students to obtain materials until a procedure, design, sketch, or some other type of evidence that supports their thought process is documented in their learning logs; once the design has been approved, instruct the materials manager to gather the needed supplies.
6. Allow students to explore, testing their hypotheses and design plans; when the CD apparatus has successfully "hovered," have students summarize their conclusions in learning logs.
7. Facilitate a discussion based on the student findings and experiences; if students failed to successfully make their craft hover, allow another student to demonstrate or provide the following instructions:
   - Place a balloon over the top of the bottle cap so that it covers the ridge on it (if there is one).
   - Blow up the balloon through the hole in the CD on the other side of the CD while holding the balloon onto the cap.
   - Pinch the balloon closed to keep the air from escaping.
   - Place the hovercraft on a flat surface, and release the balloon.
8. Challenge students to test their hovercraft over different surfaces as well as with different amounts of air.
9. Introduce the term hovercraft; have students discuss what they think causes the hovercraft to float above the surface, as well as move forward. In addition, challenge students to list things they could have done differently in order to have a more successful flight.

## Extensions for the Exploration:

- Have students measure the distance that the hovercrafts travel, as well as the time of the flight; students can then calculate speed (speed = distance ÷ time).
- Build a larger hovercraft that will hold more weight; there are many internet sites with simple directions on how to make one.
- Creative music can be used effectively to enhance the experience. If appropriate music as been selected, it can softly play while the students complete the exploration.

## Technology Connection:

An excellent site to use in conjunction with this learning experience is
www.nasaexplores.com/show2_article.php?id=03-071
The site contains the article "Two Ton Hockey Pucks," and provides further extensions for the study.

# TOTALLY TOPS

## Exploring Inquiry and Design

Time Required: At least 30 minutes for each station, plus 30-45 minutes for poster development and presentation

## How this Learning Experience Meets the National Science Education Standards:

As a result of activities in grades 5-8, all students should develop:

## Content Standard A: Science as Inquiry

### Abilities Necessary to Do Scientific Inquiry

- Identify questions that can be answered through scientific investigations.
- Design and conduct a scientific investigation.
- Use appropriate tools and techniques to gather, analyze and interpret data.
- Develop descriptions, explanations, predictions and models using evidence.
- Think critically and logically to make the relationships between evidence and explanations.
- Recognize and analyze alternative explanations and predictions.
- Communicate scientific procedures and explanations.
- Use mathematics in all aspects of scientific inquiry.

### Understanding Scientific Inquiry

- Different kinds of questions suggest different kinds of scientific investigations
- Mathematics is important in all aspects of scientific inquiry.
- Technology used to gather data enhances accuracy and allows scientists to analyze and quantify results of investigations.
- Scientific explanations emphasize evidence, have logically consistent arguments, and use scientific principles, models, and theories.
- Scientific investigations sometimes result in new ideas and phenomena for study, generate new methods or procedures for an investigation, or develop new technologies to improve the collection of data.

## Content Standard B: Physical Science

### Motions and Forces

- The motion of an object can be described by its position, direction of motion, and speed; that motion can be measured and represented on a graph.
- An object that is not being subjected to a force will continue to move at a constant speed and in a straight line.
- If more than one force acts on an object along a straight line, then the forces will reinforce or cancel one another, depending on their direction and magnitude; unbalanced forces will cause changes in the speed or direction of an object's motion.

## Content Standard E: Science and Technology

### Abilities of Technological Design:

- Design a solution or product.
- Implement a proposed design.
- Evaluate completed technological designs or products.
- Communicate the process of technological design.

### Understanding Science and Technology:

- Scientific inquiry and technological design have similarities and differences.
- Technological designs have constraints.
- Technological solutions have intended benefits and unintended consequences.

## Content Standard G: History and Nature of Science

### Science as a Human Endeavor

- Women and men of various social and ethnic backgrounds engage in activities of science, engineering, and related fields; some scientists work in teams, and some work alone, but all communicate extensively with others.
- Science requires different abilities, depending on such factors as the field and study and the type of inquiry.

### Nature of Science

- Scientists formulate and test their explanations of nature using observation, experiments, and theoretical and mathematical models.
- It is part of the scientific inquiry to evaluate the results of scientific investigations, experiments, observations, theoretical models, and the explanations proposed by other scientists.

## Objective:

Students move through three phases of inquiry as they learn about tops. Through initial learning experiences, students gain critical insight into how tops work in order to successfully design and test their own version of a top by the end of the exploration.

## Teacher Notes:

If students have not previously engaged in any type of independent learning experience where step by-step procedures are not provided, they may find the final exploratory phase frustrating; you may want to gradually ease students into full inquiry situations by providing smaller challenges periodically before engaging them in a larger experience.

Because of the large variety of supplies and materials needed for the experience, you may want to plan far enough in advance so that others can help you gather certain items. Students can also help with the collection of supplies. A "wish list" posted in the room is helpful, as well as having a "science shower" one night where parents come and help stock your supply closet.

Based on the length of class periods as well as the level of your students, you will want to consider carefully how much time to allow for each station of the rotation. Information on conducting a classroom carousel follows:

The carousel is an excellent tool to incorporate into classroom learning. The strategy serves to accomplish many goals:

- Because single stations are developed rather than 6-8, depending on the number of cooperative groups in the room, set up time is minimized for teachers with fewer materials and supplies involved.
- Students are actively engaged in their learning through the movement from station to station, a point especially important when addressing the needs of kinesthetic learners.
- Due to the time limit imposed throughout the carousel, students learn to work efficiently and effectively within a specific timeframe, a skill that becomes more and more useful as they evolve throughout their learning process and experience.
- The strategy keeps the classroom experience varied and exciting, which in turn tends to keep the students' interest piqued and therefore more focused on the tasks to be completed.
- It's fun!

Carousels can be customized to fit the needs of most classroom lessons or learning experiences. The carousel can be used during pre-assessment, engaging students, as an exploration, as part of concept development, during extensions or application phases, as an assessment tool, or any other number of ways.

### To best facilitate a carousel, the following steps are helpful:

1. Assign areas of the room that will serve as "stations" for the activity; at each station there should be a task, question, or reading to be completed.
2. Divide the students into cooperative groups or pairs.
3. Explain the following concept to the students:
   - They will have a specific amount of time to complete the task that is required at each station.
   - While at the station, they must work as a group and remain there until the signal to rotate is given.

- They are not to move ahead or try and work ahead during the activity.
- If they do not finish their task within the time allotted, the groups must still rotate to the next station when told to do so.
- Remind students to be focused and work together at each station!

4. At the end of the rotation series, each group will have completed the entire assignment given for the carousel activity.
5. Before starting, walk through the rotation if necessary to ensure that the students are aware of the direction they are to rotate. Ask the students if they have any questions concerning the procedure or activity. Announce how much time they will have at each station. Then have each of the groups go to a different station, staggering the groups if possible. Begin the activity and monitor the groups as time is kept. When the rotations are complete, have the groups return to the designated area for follow-up and discussion.

## Safety Notes:

Be sure that students handle the nails, skewers, dowels, etc., carefully. Because students will be moving around the room as they engage in the rotation between stations, be sure that the aisles and learning environment are free of obstacles; make sure that the electric pencil sharpeners are in a safe location, and no cords are extended in the walking area. Have a first aid kit available in case of cuts or puncture wounds.

## Getting Started:

1. Gather and organize the materials needed for the different phases of the learning experience.
2. Identify adequate space that will allow you to set up a carousel rotation through the three stages of the series.
3. Copy student sheets; be sure to copy front and back as needed.
4. Be prepared to effectively facilitate a carousel experience; if students are unfamiliar with the process of rotating between different stations, you may want to provide an introductory experience prior to beginning the tops series. Students must be able to work independently within a timeframe in order for the carousel to work; instructions are provided in the teacher notes.
5. Personally work through the series in order to be completely familiar with what each student should be doing and the possible outcome for each phase of the experiences.
6. Identify safety measures that may need to be discussed before students begin working with skewers, nails and other types of materials.

## Materials Needed Per Cooperative Group for the Guided Experience:

- Student sheets
- Compasses
- Masking tape
- Rubber bands
- Several pieces of poster board (5" square)
- Electric pencil sharpener
- Scissors
- 50 pennies
- Stopwatches or timers
- Small scoring pencil
- Clay

## Materials Needed Per Cooperative Group for the Challenge Experience:

- Scissors
- Rulers
- Masking tape
- Variety of pencils, bamboo skewers, dowels
- Small rubber bands
- 5 poster board sheets (9" X 12")
- Electric pencil sharpener
- Compasses
- 50 pennies
- Stopwatches or timers
- Clay
- Package of emery boards
- Box of regular paper clips

## Materials Needed Per Cooperative Group for Exploratory Experience:

Primary supply table (Basic set):

- Scissors
- Stopwatches
- Pencils
- Package of emery boards
- Round paper plates, small and large
- Clay
- Masking tape
- 50 pennies
- Skewers
- Small rubber bands
- Washer, medium and large
- Paper clips

Supplementary supply table:

- Plastic tubs (margarine, yogurt, cottage cheese)
- Foam Core
- Clay (large amount)
- Nails (variety of sizes)
- Dowels (variety of sizes)
- Toothpicks (round, flat, wood, plastic)
- Paper or plastic bowls
- Cardboard
- Ball of string
- Compasses
- Coffee stirrers

## Materials Needed Per Cooperative Group for Presentation:

- Newsprint
- Markers
- Materials from the previous tables

## Procedure:

1. While students are still in the whole group setting, outline the learning series for them; identify the location of the different stations and supply tables, as well as indicate the timeframe in which the students must work.
2. Assemble students into cooperative groups.
3. Present students with the sheet provided for the first station; review the process as needed before having the materials manager obtain the supplies and materials needed for the group.
4. Engage the students in the first station, monitoring appropriately throughout.
5. At the conclusion of the established timeframe, have students rotate to the next station.
6. Provide copies of Student Sheet #2 to each cooperative group; engage the students in the experience, monitoring as needed.
7. Following the conclusion of the second station, rotate students to the exploratory phase; provide ample time for students to fully explore tops and their behavior.
8. At the conclusion of the final station, have cooperative groups work collaboratively to design

the most effective top possible; have students base the design on what they learned and experienced as they moved through the three phases of the series.

9. Have groups prepare a newsprint poster that reflects their design along with the reasoning behind the design; if time permits, have the students construct their final top using materials from the previous experiences.
10. Allow each group to present its poster and final top, demonstrating the top as they discuss the design selected as well as the reason behind the design.
11. Facilitate a discussion throughout the presentations that highlights student observations, responses, and ideas; develop concepts that are appropriate.
12. If time permits and it is appropriate for the learning experience, conduct contests between the final tops to see which top spins longest, fastest, etc.; allow students to brainstorm about other devices they could design and test.

# TOTALLY TOPS
## Student Sheet #1

1. Using a compass and a poster board square, draw a four inch circle; cut it out.
2. Insert a small scoring pencil through the center of the circle so that the pointed end extends about ¾ of the way through the circle.
3. Push a twisted rubber band up tight against each side of the circle to stabilize the pencil so it is perpendicular to the circle.
4. Add a small piece of masking tape to further ensure that the pencil doesn't slip when it is spun.
5. Practice spinning a few times to make sure the pencil stays firmly attached and perpendicular to the circle.
6. Using masking tape, attach four evenly spaced pennies on the top surface of the top with each penny touching the pencil.
7. Take a few practice spins, and then time three spins. Record the times. (Note: A spin is considered ended when the top stops moving.)

"pennies in" Spin 1______ Spin 2______ Spin 3______

8. Now, move the four pennies to the outside edge of the top surface and attach them evenly. Re-test your top.

"pennies out" Spin 1______ Spin 2______ Spin 3______

What were your best times? Pennies in_______ Pennies out_______

What did you notice? ______________________________________________

______________________________________________

9. (If you finish early) Now add four more evenly spaced pennies. Re-test your top.

"8 pennies out" Spin 1______ Spin 2______ Spin 3______

How does this affect your times? ______________________________________________

______________________________________________

# TOTALLY TOPS
# CHALLENGING EXPERIENCE
## Student Sheet #2

1. Make a top with the spindle extending one and one-half inches below the body of the top.

2. The top must be able to spin for 10 seconds.

3. If you accomplish the first challenge, then make a top with the spindle extending three inches below the body of the top.

4. Be certain that the second top can also spin for 10 seconds.

5. For a further challenge, make another top with the spindle extending three inches below the body.

6. Have this top spin for as long as possible.

TOTALLY TOPS
Exploratory Experience
Student Sheet #3

During this experience, you are to learn as much as you can about tops. You may work alone or with a partner. A table has been made available with an assortment of materials from which you may choose. You may have access to other materials later.

Remember that you are not simply learning how to make a top. Your goal is to gain as much knowledge about tops as you can during this learning experience.

---

What I discovered about tops:

# THE INDY CARD CAR

## Putting Design into Motion

Time Required: 45-60 minutes

### How this Learning Experience Meets the National Science Education Standards:

As a result of activities in grades 5-8, all students should develop:

### Content Standard A: Science as Inquiry

Abilities Necessary to Do Scientific Inquiry

- Use appropriate tools and techniques to gather, analyze and interpret data.
- Develop descriptions, explanations, predictions and models using evidence.
- Communicate scientific procedures and explanations.
- Use mathematics in all aspects of scientific inquiry.

Understanding Scientific Inquiry

- Different kinds of questions suggest different kinds of scientific investigations.
- Mathematics is important in all aspects of scientific inquiry.
- Scientific explanations emphasize evidence, have logically consistent arguments, and use scientific principles, models, and theories.
- Scientific investigations sometimes result in new ideas and phenomena for study, generate new methods or procedures for an investigation, or develop new technologies to improve the collection of data.

### Content Standard B: Physical Science

Motions and Forces

- The motion of an object can be described by its position, direction of motion, and speed
- An object that is not being subjected to a force will continue to move at a constant speed and in a straight line.
- If more than one force acts on an object along a straight line, then the forces will reinforce or cancel one another, depending on their direction and magnitude; unbalanced forces will cause changes in the speed or direction of an object's motion.

### Content Standard E: Science and Technology

Abilities of Technological Design:

- Design a solution or product.
- Implement a proposed design.

- Evaluate completed technological designs or products.
- Communicate the process of technological design.

### Understanding Science and Technology:

- Scientific inquiry and technological design have similarities and differences.
- Technological designs have constraints.
- Technological solutions have intended benefits and unintended consequences.

## Content Standard G: History and Nature of Science

### Science as a Human Endeavor

- Women and men of various social and ethnic backgrounds engage in activities of science, engineering, and related fields; some scientists work in teams, and some work alone, but all communicate extensively with others.
- Science requires different abilities, depending on such factors as the field and study and the type of inquiry.

### Nature of Science

- Scientists formulate and test their explanations of nature using observation, experiments, and theoretical and mathematical models.
- It is part of the scientific inquiry to evaluate the results of scientific investigations, experiments, observations, theoretical models, and the explanations proposed by other scientists.

## Objective:

Using only materials provided, students work cooperatively while developing problem solving and critical thinking skills. A limited number of materials are made available therefore providing a certain level of difficulty that the students must overcome in order to complete the challenge: constructing a functional car.

## Teacher Notes:

The learning experience can be used for cooperative learning, consensus building, problem solving and/or critical thinking exclusively; or identified science concepts can be connected and developed. Examples of science content to be addressed are as follows:

- Because a ramp is used during the test, inclined planes and simple machines may be an appropriate connection.
- Calculations for speed, momentum, and other measurements associated with force and motion may be used.
- Different surfaces at the base of the ramp may be used and friction explored.

Teachers are encouraged to be creative in their use of this type of learning experience and make as many connections as possible.

Because data can be collected, it may also be appropriate to include graphing and graphical analysis at the conclusion of the learning experience.

In addition, students can be challenged at the conclusion of the experience to list what additional materials would have made the construction of the car easier for them. After writing their summary, allow students or groups of students to independently design and construct another car as an outside project.

## Safety Notes:

Students will need ample space to not only construct but test their cars. Be certain that there are no dangerous obstacles in the work area. Exercise caution when moving around the test ramp and be sure the area is clear for testing the cars. Be careful with scissors and have a first aid kit available for class use.

## Getting Started:

1. Gather the materials needed for the experience.
2. Identify additional supplies that may or may not be available for students to use.
3. Determine whether students will work in groups of two, three, or four.
4. Identify an appropriate work space for car construction, as well as an adequate place for testing the final products.
5. Construct the ramp that cars will be tested on.
6. Prepare a data table to collect class data.
7. Determine the timeframe required to complete the challenge.
8. Develop an appropriate means of assessment for the project.
9. Identify appropriate concepts to be developed using the learning experience.

## Materials Needed Per Group of Students:

- 3 X 5 index cards
- Plastic straws
- One cubic inch rubber eraser
- Tape
- Additional classroom supplies (optional)

## Materials Needed Per Class:

- Test ramp
- Measuring tool (meter stick or tape)
- Data table for class results

## The Cooperative Group Challenge:

Using the materials provided, design and construct a car that adheres to the following guidelines:

- The car must be made from only the approved materials.
- The car must be able to roll down a ramp, it's distance traveled beyond the ramp being measured.
- The car must be able to carry a one cubic rubber eraser.

## The Cooperative Group Must Perform According to These Guidelines:

- All members of the cooperative group must participate.
- The members of the group must come to a consensus on the design and construction of the car.
- The car design and construction must be completed within the timeframe stated.
- The group's car must be tested in front of the whole group, passing the same test as all other cars.

## Procedure:

1. Arrange students into cooperative teams; assign tasks within the groups as appropriate.
2. Provide each group with the design challenge.
3. Have materials managers gather the needed materials.
4. Initiate the learning experience.
5. Following the time limit, assess each car by rolling it down the test ramp and measuring the distance traveled beyond the end of the ramp.
6. Facilitate a discussion and develop concepts as needed.

# THE INDY CARD CAR
## The Cooperative Group Challenge

Using the materials provided, design and construct a car that adheres to the following guidelines:

- The car must be made from only the approved materials.
- The car must be able to roll down a ramp, it's distance traveled beyond the ramp being measured.
- The car must be able to carry a one cubic rubber eraser.

The cooperative group must perform according to these guidelines:

- All members of the cooperative group must participate.
- The members of the group must come to a consensus on the design and construction of the car.
- The car design and construction must be completed within the timeframe stated.
- The group's car must be tested in front of the whole group, passing the same test as all other cars.

THE CAR MEASURING THE GREATEST DISTANCE WINS THE CHALLENGE!

# THE CARDBOARD CHAIR

## A Challenge to Create a Chair

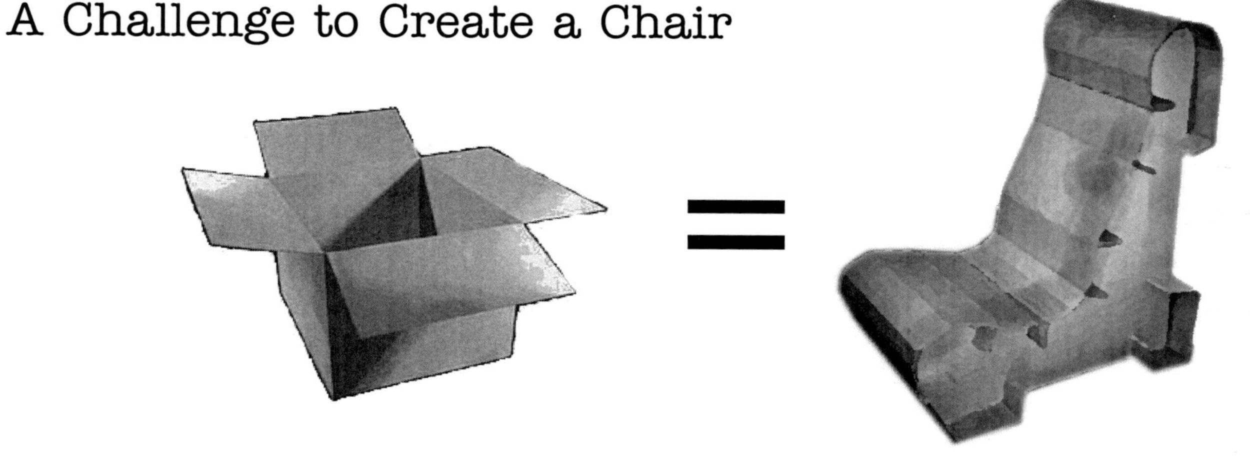

Time Required: 60-90 minutes plus discussion

## How this Learning Experience Meets the National Science Education Standards:

As a result of activities in grades 5-8, all students should develop:

### Content Standard E: Science and Technology

Abilities of Technological Design:

- Design a solution or product.
- Implement a proposed design.
- Evaluate completed technological designs or products.
- Communicate the process of technological design.

Understanding Science and Technology:

- Scientific inquiry and technological design have similarities and differences.
- Technological designs have constraints.
- Technological solutions have intended benefits and unintended consequences.

### Content Standard G: History and Nature of Science

Science as a Human Endeavor

- Women and men of various social and ethnic backgrounds engage in activities of science, engineering, and related fields; some scientists work in teams, and some work alone, but all communicate extensively with others.
- Science requires different abilities, depending on such factors as the field and study and the type of inquiry.

Nature of Science

- Scientists formulate and test their explanations of nature using observation, experiments, and theoretical and mathematical models.
- It is part of the scientific inquiry to evaluate the results of scientific investigations, experiments, observations, theoretical models, and the explanations proposed by other scientists.

## Background:

It has been said that the most positive aspect of utilizing design technology projects is that students and their teachers begin to look at problems and issues from multiple points of view and in relationship to a variety of contexts. One problem may even create another problem and there are usually several different solutions to each problem.

Students also learn that design technology, or engineering design, like life itself, is an endless process of solving problems. In solving any problem, people take the same steps as the ones that students will utilize in their design technology experiences:

- Stating the problem clearly.
- Collecting information.
- Developing possible solutions.
- Selecting the best solutions.
- Implementing the solution.
- Evaluating the solution.
- Making the needed changes and improving the solution.
- Communicating their findings.

The action of solving problems also opens up the creative process for students, thus enhancing the engagement of students in the classroom learning. Reports indicate that when students are building and creating things in the classroom, the engagement level is consistently intense. It does not allow a student to simply sit back and wait to be told what to do, but instead requires that the student create, test, and evaluate for themselves. This in turn leads to genuine decision making, which should be an integral part of the entire curriculum. The goal of problem solving is to educate students to be able to use the scientific process no matter what the problem is that they encounter.

## Teacher Notes:

The following concepts can be developed fully through an exploration in engineering design:

- Forces
- Torque
- Center of mass
- Stability
- Tension
- Compression
- Beam strength
- Column strength

An excellent reference to use with developing learning experiences related to these concepts is The Exploratorium Guide to Scale and Structure, Barry Kluger-Bell and the School in the Exploratorium, Heinemann, a division of Redd Elsevier, 1995.

## Safety Notes:

Participants should be reminded of the safe handling of cutting implements; caution should be exercised when cutting the cardboard pieces. Have a first aid kit available. Students should be careful when testing the chairs—it may be advisable to test the chairs with substitute weight before having an individual sit in the chair to be certain it can support the person. In addition, the work area should be clear of obstacles. Movement throughout the work area should be careful.

## Getting Started:

1. Collect the abundance of cardboard needed for the challenge; you may want to have appliance stores, discount houses, etc. begin saving large boxes for you in advance of the project.
2. Gather implements for cutting cardboard.
3. Determine the size of the groups engaging in the experience.
4. Identify a work area that can accommodate the construction and testing of the cardboard chairs.
5. Copy student challenge sheets.
6. Copy Engineering Design checklists, if needed.

## Materials Needed Per Group of Students:

- Cardboard
- Implements for cutting cardboard
- Student challenge sheets
- Engineering Design checklists (if appropriate)

## Procedure:

1. Assemble students into cooperative groups.
2. Challenge each group to construct a cardboard chair within the following guidelines:
    - Each group has 60-90 minutes to complete construction and testing of their chair.
    - All chairs must be made from cardboard only—no tape, glue, staples, etc.
    - The structure must resemble a chair.
    - Chairs must be free standing and not attached to anything for support; a footstool made from cardboard and attached to the chair will be allowed.
    - Each group must be able to pick up its chair as one piece.
    - The chair must support the weight of one of its group members.
    - The chair that can hold the most weight will wins the challenge.

# Cardboard Chair
# Student Sheet

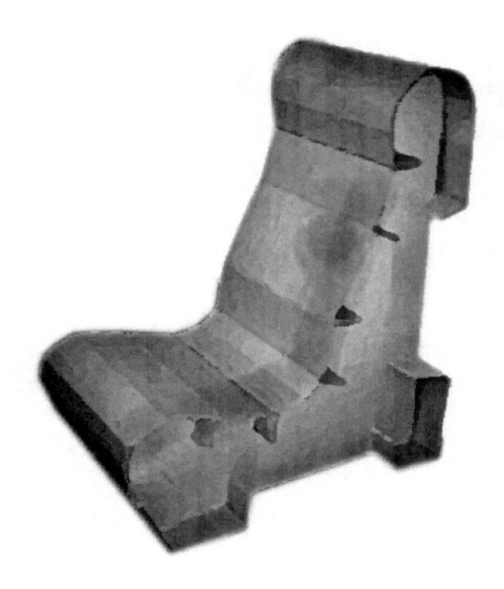

## Challenge Guidelines:

Construct a cardboard chair within the following guidelines:

- Each group has 60-90 minutes to complete construction and testing of their chair.
- All chairs must be made from cardboard only—no tape, glue, staples, etc.
- The structure must resemble a chair.
- Chairs must be free standing and not attached to anything for support; a footstool made from cardboard and attached to the chair will be allowed.
- Each group must be able to pick up its chair as one piece.
- The chair must support the weight of one of its group members.
- The chair that can hold the most weight will wins the challenge.

# Engineering Design Checklist
# Student Sheet

______ The structure meets the stated challenge requirements.

______ The project design team followed the challenge rules.

______ The project team exhibited effective, cooperative group work with every member participating.

______ The project prototype was carefully designed and built (if required).

______ The project team showed perseverance and a willingness to try again if necessary.

______ The project team was inventive in redesigning the original prototype.

______ The final structure shows creativity and originality.

______ The project meets the stated goals of the team.

# Mechanical Engineering Word Search Puzzle

I R S F H G K Y F U J X E D D S J I C R G Y L M E
L H K F Q H X G N R T W K Q B S Y S T E M S S W J
K R Q T E R N D G N I D N A T S T U O C C S I T G
Z Y B F E H E Q G U J I E S B R Z H H A Z T N T K
T G F T Q R H W E E M W N B A O O T B R K Y K S Y
X R B Y W G N D O S I B V L G P B U S C O J U Q J
Q E V A Q U I T X P N A I Y S R J U V E J T Z M Y
M N T T M V E C A P S I R E C A S R L M D R R O G
F E X J E E K J L T O M O G S U P G U I D A N C E
R B J R X Y D J B P U P N A J T O M R Y V T O B P
C C S Y C H U I H G Z R M T V O R T R B K S B R I
S E X D E Q C K C V C O E N E M T Y C J E L F V B
H Z R E L F E U L A S V N A H O S A F U W P Q M T
P B M J L F D B O D L E T V I T O S M L D H T Z D
P K Y V E Y T I L A U Q A D C I U M J Z V O K G Z
Y W B L N T T M S O B O L A L V I E Q R O C R Q Z
L A M M T Y T C I F F A R T E E D N H W J H H P D

1. A Larger population = more jobs in ___________ pollution.
2. ME's design __________ structures.
3. Some schools offer ME with an emphasis in _________ engineering.
4. According to the ASME, the job opportunitie for mechanical engineers are _____________.
5. ME's make _________ money.
6. ME student magazine: The Mechanical _____________.
7. ME is oene of the _________ disciplines.
8. ME is a very ___________ discipline.
9. ME's are needed to develop vehicle _________ systems.
10. ME's design transportation __________.
11. ME's are involved in the miniaturization of _________ instruments.
12. ME's design and develop every kind of __________.
13. ME's are needed to develop ___________ technologies.
14. ME's ___________ air quality by reducing pollution.
15. College is the best time to _________ a business because of all the available resources.
16. Most machines are the _________ of an ME
17. ME's help to develop ________ equipment.
18. ME's develop structures for _________ travel.
19. ME's help reduce _______ congestion.
20. American Society of Mechanical Engineers.
21. ME's work in __________ assurance.
22. ME's manufacture ___________ systems.
23. ME's design _______ engines.

# Mechanical Engineering Word Search Puzzle
## Teacher Sheet

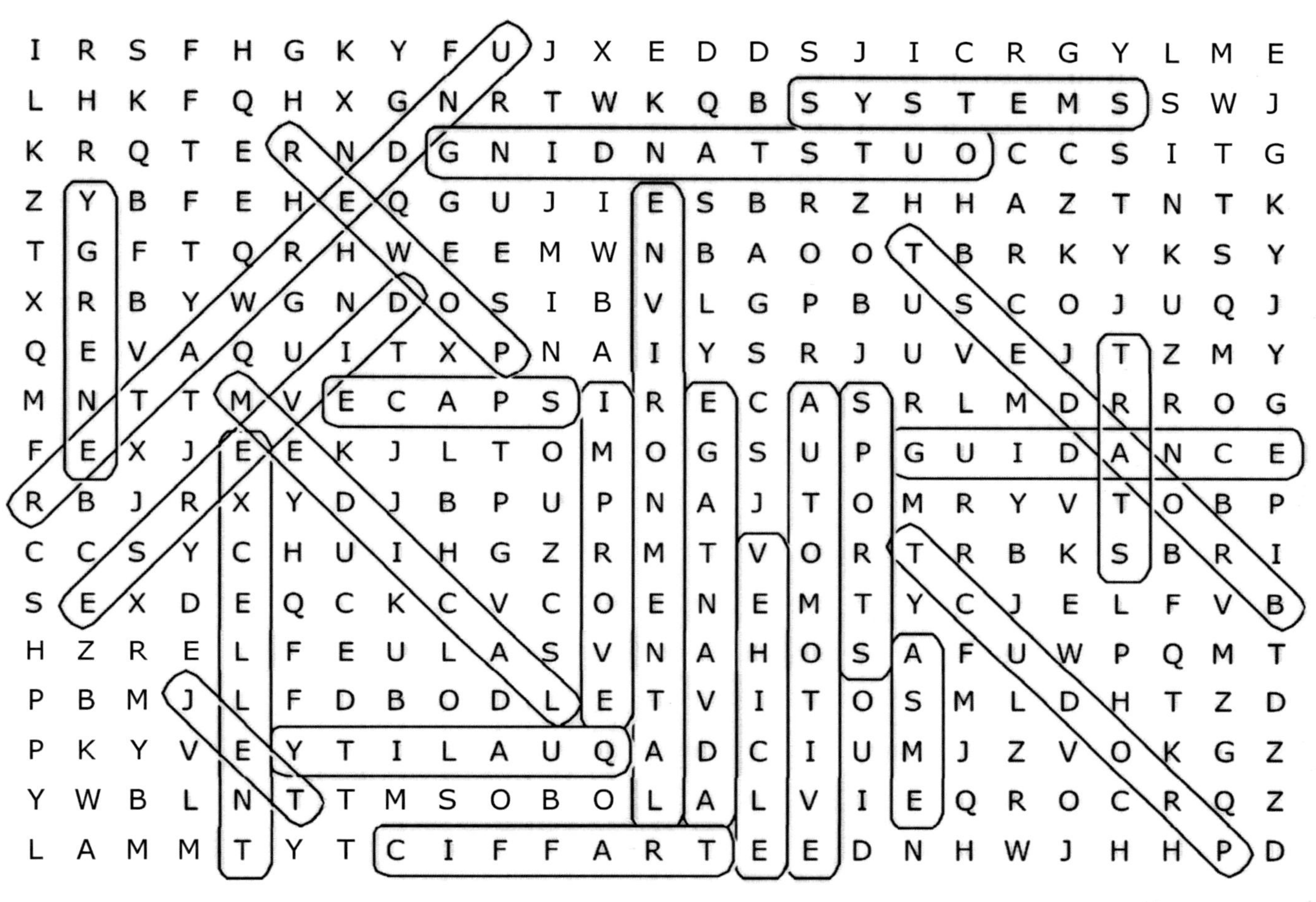


1. A Larger population = more jobs in ___________ pollution. [environmental]

2. ME's design __________ structures. [underwater]

3. Some schools offer ME with an emphasis in _________ engineering. [automotive]

4. According to the ASME, the job opportunities for mechanical engineers are _____________. [outstanding]

5. ME's make _________ money. [excellent]

6. ME student magazine: The Mechanical ____________. [advantage]

7. ME is oene of the _________ disciplines. [broadest]

8. ME is a very ___________ discipline. [diverse]

9. ME's are needed to develop vehicle _________ systems. [guidance]

10. ME's design transportation __________. [systems]

11. ME's are involved in the miniaturization of _________ instruments. [medical]

12. ME's design and develop every kind of __________. [vehicle]

13. ME's are needed to develop ___________ technologies. [energy]

14. ME's ___________ air quality by reducing pollution. [improve]

15. College is the best time to _________ a business because of all the available resources. [start]

16. Most machines are the _________ of an ME. [product]

17. ME's help to develop ________ equipment. [sports]

18. ME's develop structures for _________ travel. [space]

# Mechanical Engineering Crossword Puzzle

**Across**

1. ME's are needed to develop vehicle _________ systems.
4. ME's ___________ air quality by reducing pollution.
6. ME is a very ___________ discipline.
8. ME's develop structures for _________ travel.
10. College is the best time to _________ a business because of all the available resources.
11. ME's design _______ engines.
12. ME is oene of the _________ disciplines.
14. ME's design __________ structures.
15. ME's make _________ money.
18. ME's help to develop ________ equipment.
21. ME's help reduce _______ congestion.
22. ME's design and develop every kind of _________.

**Down**

2. ME student magazine: The Mechanical ____________.
3. ME's manufacture ___________ systems.
5. According to the ASME, the job opportunities for mechanical engineers are _____________.
7. ME's are involved in the miniaturization of _________ instruments.
9. A Larger population = more jobs in ___________ pollution.
10. ME's design transportation __________.
13. Some schools offer ME with an emphasis in _________ engineering.
16. ME's are needed to develop ___________ technologies.
17. ME's work in __________ assurance.
19. Most machines are the _________ of an ME.
20. American Society of Mechanical Engineers.

# Mechanical Engineering Crossword Puzzle
## Teacher Sheet

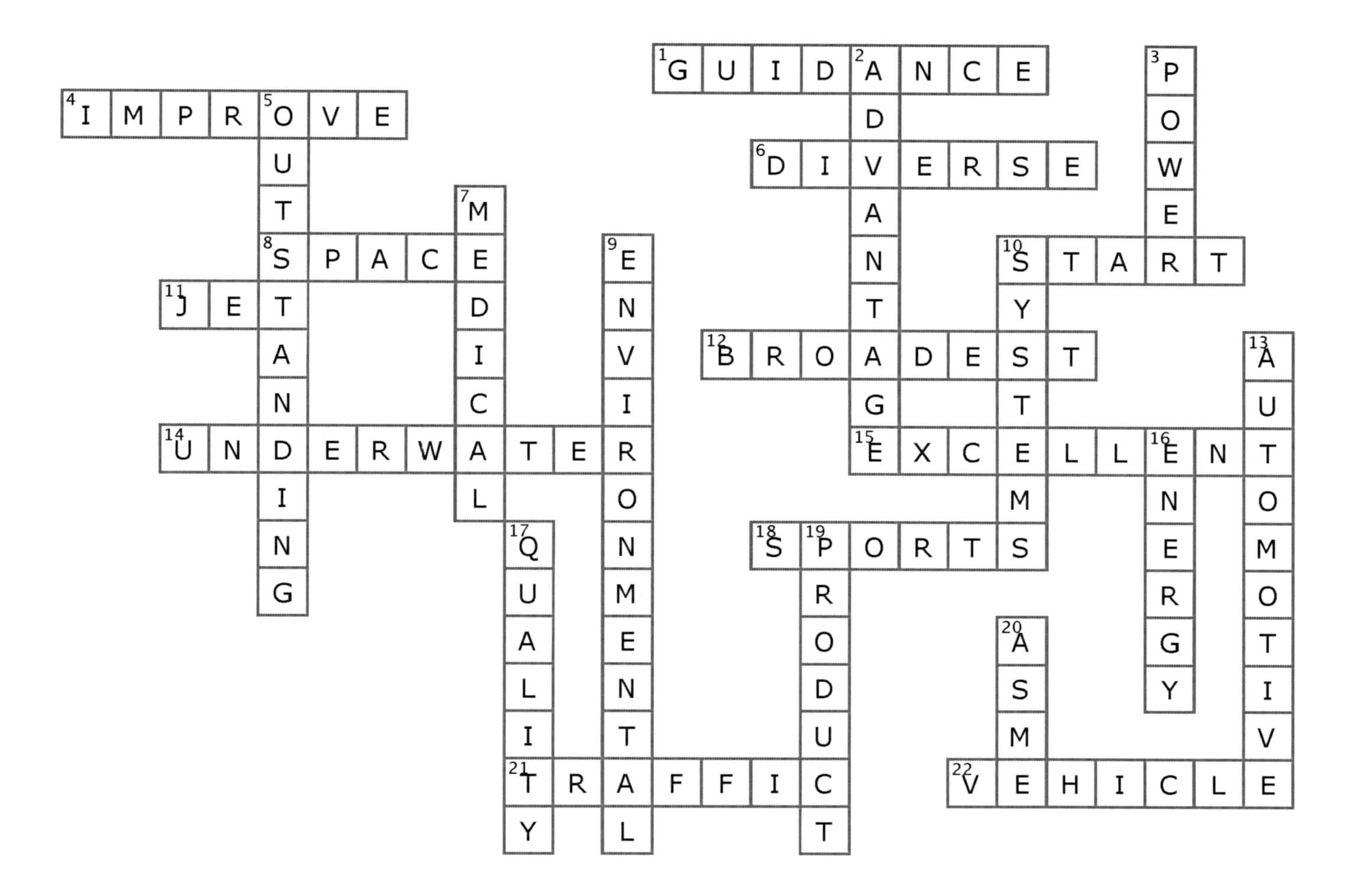


**Across**

1. ME's are needed to develop vehicle _________ systems. [guidance]
4. ME's ___________ air quality by reducing pollution. [improve]
6. ME is a very ___________ discipline. [diverse]
8. ME's develop structures for _________ travel. [space]
10. College is the best time to _________ a business because of all the available resources. [start]
11. ME's design _______ engines. [jet]
12. ME is oene of the _________ disciplines. [broadest]
14. ME's design __________ structures. [underwater]
15. ME's make _________ money. [excellent]
18. ME's help to develop ________ equipment. [sports]

**Down**

2. ME student magazine: The Mechanical ____________. [advantage]
3. ME's manufacture ___________ systems. [power]
5. According to the ASME, the job opportunities for mechanical engineers are _____________. [outstanding]
7. ME's are involved in the miniaturization of _________ instruments. [medical]
9. A Larger population = more jobs in ___________ pollution. [environmental]
10. ME's design transportation __________. [systems]
13. Some schools offer ME with an emphasis in _________ engineering. [automotive]
16. ME's are needed to develop ___________ technologies. [energy]
17. ME's work in __________ assurance. [quality]

# Civil Engineering

## Median Income: $74,000

Civil engineering is one of the oldest and largest branches of engineering. Traditionally, civil engineers planned and designed such things as roads, bridges, high-rises, dams, and airports. Because of population growth and a booming economy, however, the civil engineer now designs new things such as underwater tunnels, new and better wastewater treatment plants, solutions for highway congestion, and special tracks for the magnetic levitation trains of the future. As our current understanding of technology increases, demand for the diverse talents of civil engineers will increase too.

According to the Discover Engineering Web site at www.discoverengineering.org, civil engineering is "the art and science of designing the infrastructure of a modern CIVILized society." Money magazine ranked civil engineering the tenth highest paying profession and the fifth best job in America out of 100, based on salary, prestige, growth and security in 1994.

Civil engineers enjoy being able to choose from many specialties within their discipline such as environmental, transportation, structural, geotechnical, or water/wastewater management.

Students who pick a track in:

- Environmental engineering might be interested in monitoring air pollution or transforming wetlands into golf courses.
- Geotechnical engineering may help maintain the Hoover dam or construct seawalls.
- Structural engineering might be interested in working to make buildings or roads earthquake safe. The structural engineering student may also design offshore oilrigs or even a coliseum that hosts sporting events and concerts.
- Transportation engineering might design the transportation of the future. That may include working with high-speed trains or new types of boats.

There is no limit to the versatility and opportunity of the civil engineering profession.

Approximately 60 percent of civil engineers work in construction, transportation, manufacturing and utilities. Most of the remaining 40 percent work for federal, state, or local government agencies. Almost half of these governmental jobs involve developing designs as engineering consultants.

Some civil engineers opt for a career in research where they may attempt to find newer, stronger, and more resilient materials. A civil engineer who goes to law school could be an earthquake or hurricane insurance attorney, or simply a specialized attorney for a major construction company.

The American Society of Civil Engineers (ASCE) offers a student chapter to help students step easily into the real world of civil engineering. The organization has 15,000 student members across the nation. ASCE supports students and encourages growth by offering numerous scholarships, competitions and awards to its student members. Virtual Student Community, a student-dedicated site, gives you access to information that will help smooth the path to your career in civil engineering. This access includes a database of job postings, student chapter links, event calendars, membership contests, membership forms, discussion forums dedicated to topics that affect civil engineering students, and much more. For more information, visit the ASCE Web site at www.asce.org.

## Civil Engineers and the Things They Design:

### Skateboards:
They may design skate parks that allow the skaters to jump benches, slide down poles, ledges and or invent other obstacles that allows the skaters to do tricks. Civil engineers may also design portable ramps and ledges for learning to do tricks at home, or to allow the skater to setup a make-shift park anywhere they want.

### Pools and equipment for swimmers:
They may design swimming pools that reduce waves for competitions. Some may also design heating systems that always keep the pool at a constant temperature.

### Tennis:
Materials engineers may work with civil engineers to design new courts surfaces that reduce injury while also slowing or speeding up the game.

### Bowling:
May design the bowling alley itself.

### Roller Coasters
They may design the framing and structure.

### Pollution Doctors
They may figure out what is wrong with the air or water and make it better. They are also known as environmental engineers.

### Water Cleaners
They may dedicate themselves to making facilities to clean water for drinking or taking the "waste" out of wastewater. Water resource engineers figure out ways of collecting ample water for a community.

### Ground Detectives
They may know all about the earth's crust and make sure anything going into the ground or sitting on the ground (which is pretty much everything) is safe and will stay where they were intended.

### Water Detectives
They may know all about the rivers, lakes and oceans, and know how to best make it possible for boats to travel in them, or dams to harness their energy or restrain them from flooding communities.

### Transportation Specialists
They may make transportation systems telling people when to stop, when to go, and where to go, preventing accidents and making travel quick and smooth.

### Road Experts
They may know all about roads, knowing where best they should go, what it takes to make them happen, how to make them safe, and how to connect them to each other.

### Land Experts
They may be experts in figuring out all the shapes and contours of a piece of land, and know how to use or modify those shapes to make a great new place.

### Construction Experts

They may know how to plan and execute a construction job to get it done right and done fast, and dedicate themselves to making sure that happens.

### Building Materials Experts

They may understand steel, concrete, wood, and all sorts of other materials, and how to make structures like buildings, bridges and tunnels using them.

All civil engineers use all they have learned, and follow the rules of a community to create a solution to a problem that will make everyday life better for the people.

## Reed Brockman
## Structural Engineer

My special interest is in bridge inspection and rehabilitation. I look at each bridge as if it were a patient – a big patient. Some patients look healthy when you first visit them, but turn out to have something wrong, while some others that look beat up are in generally good health. Either way, I take pride in knowing what symptoms to expect, and what should be done to get the bridges back into shape. Actually, there's a great many other things I enjoy about my work, such as: getting outdoors, climbing, constantly using my communication skills, and solving the puzzle of accessing some of the more complicated structures (renting cranes, barges, snooper trucks, etc.).

I equally enjoy designing bridges and tunnels, especially coordinating the many different aspects into one design: coordinating the needs of the neighbors, maintaining overhead clearances, making sure rainwater can find its way off the roadway, giving all the pipes and cable ducts a home, and of course making it strong, easily maintained and beautiful.

When I was in high school, I had no idea what I was going to become. I was a good student (mathlete, yearbook editor, and some other things), and liked English and History just as much as Math and Science. Outside school, I was obsessed with music (and still am – I am also a DJ). I think I had memorized every word of every local punk band... With my head pulling me in so many directions, I had no idea on what I would like to focus when I was in college.

When I was applying for schools, I tried to look at Universities that had a wide variety of choices. In general, I applied to become an engineering student, mostly because my brothers were both engineering students and they seemed happy enough. I chose the University of Pennsylvania, knowing they had great professors in most courses of study in which I was interested. My first semester, I majored in electrical engineering, but knew I was totally unconvinced that I wanted to actually be one. After one semester, I figured out civil engineering and architecture were more my style, and decided to take on both.

The summer after my junior year I got a job working as a bridge inspector with a firm named Barnes & Jarnis, and decided at the end of the summer that I wanted to work there after graduation. This meant abandoning architecture and not even staying on for a masters. This also meant I wouldn't be staying on to get a Masters in Civil Engineering. I have moments I regret this, but they are infrequent: I've always enjoyed my work.

Where I've been lucky is in having great mentors. In high school, one teacher (Marty Badoian) went well beyond the classroom to encourage me to give my all in everything I do. In college, one professor (Rick Johansen) opened my eyes to the real world of engineering beyond computations. And in my early days working, I was very spoiled. Between Ralph Salvucci, Ted Long and Darren Conboy, I always knew where to turn for advice.

Speaking of advice, here's some for you. If anyone tells you to only consider engineering if you love math and science, ignore them. If you love to pool everything you know together and solve problems, and then see real results, that is the civil engineering I've always known. I use what I learned as yearbook editor every bit as much as physics or calculus.

# BOLSTERING BOOKS

## Producing a Paper Platform

Time Required: 30 minutes plus discussion

### How this Learning Experience Meets the National Science Education Standards:

As a result of activities in grades 5-8, all students should develop:

### Content Standard E: Science and Technology

Abilities of Technological Design:

- Design a solution or product.
- Implement a proposed design.
- Evaluate completed technological designs or products.
- Communicate the process of technological design.

Understanding Science and Technology:

- Scientific inquiry and technological design have similarities and differences.
- Technological designs have constraints.
- Technological solutions have intended benefits and unintended consequences.

### Content Standard G: History and Nature of Science

Science as a Human Endeavor

- Women and men of various social and ethnic backgrounds engage in activities of science, engineering, and related fields; some scientists work in teams, and some work alone, but all communicate extensively with others.
- Science requires different abilities, depending on such factors as the field and study and the type of inquiry.

Nature of Science

- Scientists formulate and test their explanations of nature using observation, experiments, and theoretical and mathematical models.
- It is part of the scientific inquiry to evaluate the results of scientific investigations, experiments, observations, theoretical models, and the explanations proposed by other scientists.

## Background:

It has been said that the most positive aspect of utilizing design technology projects is that students and their teachers begin to look at problems and issues from multiple points of view and in relationship to a variety of contexts. One problem may even create another problem and there are usually several different solutions to each problem.

Students also learn that design technology, or engineering design, like life itself, is an endless process of solving problems. In solving any problem, people take the same steps as the ones that students will utilize in their design technology experiences:

- Stating the problem clearly.
- Collecting information.
- Developing possible solutions.
- Selecting the best solutions.
- Implementing the solution.
- Evaluating the solution.
- Making the needed changes and improving the solution.
- Communicating their findings.

The action of solving problems also opens up the creative process for students, thus enhancing the engagement of students in the classroom learning. Reports indicate that when students are building and creating things in the classroom, the engagement level is consistently intense. It does not allow a student to simply sit back and wait to be told what to do, but instead requires that the student create, test, and evaluate for themselves. This in turn leads to genuine decision making, which should be an integral part of the entire curriculum. The goal of problem solving is to educate students to be able to use the scientific process no matter what the problem is that they encounter.

## Teacher Notes:

The following concepts can be developed fully through an exploration in engineering design:

- Forces
- Torque
- Center of mass
- Stability
- Tension
- Compression
- Beam strength
- Column strength

An excellent reference to use with developing learning experiences related to these concepts is The Exploratorium Guide to Scale and Structure, Barry Kluger-Bell and the School in the Exploratorium, Heinemann, a division of Redd Elsevier, 1995.

## Safety Notes:

Participants should be reminded of the safe handling of scissors; in addition, the work area should be clear of obstacles. Movement throughout the work area should be careful.

## Getting Started:

1. Gather materials.
2. Determine the size of the groups engaging in the experience.
3. Identify a work area that can accommodate the construction and testing of the book platforms.
4. Copy student challenge sheets.
5. Copy Engineering Design checklists, if needed.

## Materials Needed Per Group of Students:

- A variety of books (one per group)
- 3 sheets of white paper
- 1 meter of masking tape
- 2 pair of scissors
- 2 pencils
- Student challenge sheets
- Engineering design checklists (if appropriate)

## Procedure:

1. Assemble students into cooperative groups.
2. Challenge each group to construct a paper tube platform that will hold a book six inches off the table; the following guidelines should be followed:
    - Each group has 30 minutes to complete construction and testing of their platform.
    - All tubes must be formed by rolling the paper over pencils, then removing the pencils.
    - Tubes can be taped together.
    - No more than twenty tubes can be used.
    - Construct the final tower from the construction paper.
    - Groups will get no more tape or paper during the challenge.
    - The platform must be able to completely support the chosen book for twenty seconds.
    - The paper platform that supports the heaviest book at least six inches above the table for twenty seconds wins the challenge.

# Student Sheet
# Challenge Guidelines

Construct a paper platform that will support a book six inches off the table within the following guidelines:

- Each group has 30 minutes to complete construction and testing of their platform.

- All tubes must be formed by rolling the paper over pencils, then removing the pencils.

- Tubes can be taped together.

- No more than twenty tubes can be used.

- Construct the final tower from the construction paper.

- Groups will get no more tape or paper during the challenge.

- The platform must be able to completely support the chosen book for twenty seconds.

- The paper platform that supports the heaviest book at least six inches above the table for twenty seconds wins the challenge.

# Engineering Design Checklist
# Student Sheet

______ The structure meets the stated challenge requirements.

______ The project design team followed the challenge rules.

______ The project team exhibited effective, cooperative group work with every member participating.

______ The project prototype was carefully designed and built (if required).

______ The project team showed perseverance and a willingness to try again if necessary.

______ The project team was inventive in redesigning the original prototype.

______ The final structure shows creativity and originality.

______ The project meets the stated goals of the team.

# Pasta Bridges

## An Edible Edifice

**What allows an arch bridge to span greater distances than a beam bridge, or a suspension bridge to span a distance seven times that of an arch bridge?** ***The answer lies in how each bridge type deals with two important forces called compression and tension.***
Reprinted from: Howstuffworks.com

Time Required: 60 minutes plus discussion

## How this Learning Experience Meets the National Science Education Standards:

As a result of activities in grades 5-8, all students should develop:

### Content Standard E: Science and Technology

Abilities of Technological Design:

- Design a solution or product.
- Implement a proposed design.
- Evaluate completed technological designs or products.
- Communicate the process of technological design.

Understanding Science and Technology:

- Scientific inquiry and technological design have similarities and differences.
- Technological designs have constraints.
- Technological solutions have intended benefits and unintended consequences.

### Content Standard G: History and Nature of Science

Science as a Human Endeavor

- Women and men of various social and ethnic backgrounds engage in activities of science, engineering, and related fields; some scientists work in teams, and some work alone, but all communicate extensively with others.
- Science requires different abilities, depending on such factors as the field and study and the type of inquiry.

Nature of Science

- Scientists formulate and test their explanations of nature using observation, experiments, and theoretical and mathematical models.
- It is part of the scientific inquiry to evaluate the results of scientific investigations, experiments, observations, theoretical models, and the explanations proposed by other scientists.

# Background:

A bridge is a structure in which most of the structure's weight is out and away from the points of support. Building bridges is more of a challenge than building an upright structure because most of the weight being supported is not directly over a base. Building bridges involves students in direct experiences with the concepts of torque and center of mass. Torque is a force that is equal to the combination of the downward force (weight) of an object and the distance of the object from the pivot point (torque = force X distance). Torques test to twist or change the state of rotation of things. For a bridge, the greatest torque occurs at the midpoint of the bridge. The longer the bridge, the greater the torque when equal downward forces (weights) are exerted on it.

The center of mass is the balance point for any object. For a given body, the center of mass is the average position of all the particles of mass that constitute the object. We can balance an object by applying an upward force equal to its weight at the object's center of mass. The center of mass of an object may be at a point where no mass exists; the center of mass of a ring or hollow sphere is at the geometrical center where no mass exists. The location of the center of mass is important to the stability of a structure. If a bridge is a symmetrical structure, its center of mass is at its midpoint.

Two other forces that must be considered when building any kind of structure are tension and compression. Tension occurs when a material is being pulled apart; tension lengthens materials.

Compression occurs when materials are pushed together; compression shortens materials. Compression and tension are opposite forces. Tension and compression must be balanced if a structure is to maintain its stability.

If a load is placed on a beam or a bridge that is fastened at two ends, the beam or bridge experiences both tension and compression. The top side of the bridge experiences a very small amount of compression from the weight of the load, while the bottom side experiences the comparable tension.

The middle axis of the beam is not stressed at all. When a load is placed on a beam, most of the force of tension and compression falls at the farthest points from the middle axis. This is why tubes are so strong; most of their material lies on the edges of the tube, precisely where most of the forces are exerted.

# Challenges for Bridge Builders:

Many materials are suitable for building bridges, including pasta, straws, clay and paper. Limited materials should be provided to students; for example, fifty straws and fifty straight pins. Using materials provided, students can be challenged to:

- Build a bridge from one desk or table to another before determining how much weight it will hold.
- Build the longest bridge possible with the given materials.
- Build the strongest bridge possible that will cover a given distance.

Students will have many questions when building bridges about the "rules" for bridge building. It may be necessary to define certain rules ahead of time. For example, the ends of the bridges cannot be anchored to their support and intermediate supports cannot touch the ground. Students should probably discuss and decide as a group where on a bridge weights should be hung to test the strength of a bridge.

# Extensions

- Wind-test structures by using a large piece of cardboard as a fan or using an electric fan.
- Have students search for unnecessary pieces by one at a time cutting or removing pieces, then seeing if their structure remains intact. If pieces are not necessary, challenge students to build a structure that will have no unnecessary pieces. In real life, the cost of construction materials necessitates using the least amount of material that will still provide a safe structure.
- Provide students with materials that they can use with their families to build a structure. Have students take a picture or make a drawing of the structure they build with their families and write about the process.

# Assessing Student Learning:

Design construction provides a perfect situation for involving students in a performance assessment.

- Provide students with new materials or a slightly different challenge from the one they have been working on, and note how well they incorporate the concepts they have learned into their final structure.
- Ask students to write about the process to determine their understanding of basic construction concepts such as torque, center of mass, stability, etc.
- Another assessment technique would be to provide students with a model or picture of a structure and ask them what would happen if weights were placed in specific locations or if certain structural units were removed.

# Teacher Notes:

The following concepts can be developed fully through an exploration in engineering design:

- Forces
- Torque
- Center of mass
- Stability
- Tension
- Compression
- Beam strength
- Column strength

An excellent reference to use with developing learning experiences related to these concepts is The Exploratorium Guide to Scale and Structure, Barry Kluger-Bell and the School in the Exploratorium, Heinemann, a division of Redd Elsevier, 1995.

# Safety Notes:

Participants should be reminded of the safe handling of scissors; in addition, the work area should be clear of obstacles. Movement throughout the work area should be careful.

# Getting Started:

1. Collect the materials needed for the challenge.
2. Determine the size of the groups engaging in the experience.
3. Identify a work area that can accommodate the construction and testing of the pasta bridges.
4. Copy student challenge sheets.
5. Copy Engineering Design checklists, if needed.

## Materials Needed Per Group of Students:

- 1 package uncooked pasta (spaghetti or fettucine)
- 1 meter masking tape
- Scissors
- Weights, washers, or film canisters filled with sand
- Large paper clips to attach weights
- Meter stick
- Balance or scale (if needed to determine weight of washers/film canisters)
- Student challenge sheets
- Engineering design checklists (if appropriate)

## Procedure:

1. Assemble students into cooperative groups.
2. Challenge each group to construct the strongest pasta bridge within the following guidelines:
   - Each group has 60 minutes to complete construction and testing of their bridge.
   - All bridges must be at least one meter in length.
   - The ends of the bridge cannot be attached to the table or desk in any way.
   - Intermediate supports of the bridge cannot touch the ground.
   - Weights must be hung from the center of the bridge when testing.
   - Groups will get no more pasta or tape during the challenge.
   - The bridge that can support the most weight will wins the challenge.

For bridge strength and building ideas, do an Internet search for various types of trusses such as: Warren, Pratt and Howe.

# Pasta Bridges
# Student Sheet

## Challenge Guidelines:

Construct a strong pasta bridge within the following guidelines:

- Each group has 60 minutes to complete construction and testing of their bridge

- All bridges must be at least one meter in length.

- The ends of the bridge cannot be attached to the table or desk in any way.

- Intermediate supports of the bridge cannot touch the ground.

- Weights must be hung from the center of the bridge when testing.

- Groups will get no more pasta or tape during the challenge.

- The bridge that can support the most weight will wins the challenge.

# Engineering Design Checklist
# Student Sheet

______ The structure meets the stated challenge requirements.

______ The project design team followed the challenge rules.

______ The project team exhibited effective, cooperative group work with every member participating.

______ The project prototype was carefully designed and built (if required).

______ The project team showed perseverance and a willingness to try again if necessary.

______ The project team was inventive in redesigning the original prototype.

______ The final structure shows creativity and originality.

______ The project meets the stated goals of the team.

# Engineering Design Project
# Student Report

________________________________________

Name of Structure Designed and Built

| OBSERVATIONS: | QUESTIONS: |
|---|---|
| | |

Drawings and Reflections:

# Ten Second Tower

## A Construction Paper Construction Challenge

Time Required: 30 minutes plus discussion

### How this Learning Experience Meets the National Science Education Standards:

As a result of activities in grades 5-8, all students should develop:

### Content Standard E: Science and Technology

Abilities of Technological Design:

- Design a solution or product.
- Implement a proposed design.
- Evaluate completed technological designs or products.
- Communicate the process of technological design.

Understanding Science and Technology:

- Scientific inquiry and technological design have similarities and differences.
- Technological designs have constraints.
- Technological solutions have intended benefits and unintended consequences.

### Content Standard G: History and Nature of Science

Science as a Human Endeavor

- Women and men of various social and ethnic backgrounds engage in activities of science, engineering, and related fields; some scientists work in teams, and some work alone, but all communicate extensively with others.
- Science requires different abilities, depending on such factors as the field and study and the type of inquiry.

Nature of Science

- Scientists formulate and test their explanations of nature using observation, experiments, and theoretical and mathematical models.
- It is part of the scientific inquiry to evaluate the results of scientific investigations, experiments, observations, theoretical models, and the explanations proposed by other scientists.

## Teacher Notes:

The following concepts can be developed fully through an exploration in engineering design:

- Forces
- Torque
- Center of mass
- Stability
- Tension
- Compression
- Beam strength
- Column strength

## Safety Notes:

Participants should be reminded of the safe handling of scissors; in addition, the work area should be clear of obstacles. Movement throughout the work area should be careful.

## Getting Started:

1. Collect the materials needed for the challenge.
2. Determine the size of the groups engaging in the experience.
3. Identify a work area that can accommodate the construction and testing of the paper towers.
4. Copy student challenge sheets.
5. Copy Engineering Design checklists, if needed.

## Materials Needed Per Group of Students:

- 1 piece of construction paper
- 1 piece of white paper
- 1 meter masking tape
- Scissors
- Student challenge sheets
- Engineering design checklists (if appropriate)

## Procedure:

1. Assemble students into cooperative groups.
2. Challenge each group to construct the tallest free-standing tower possible from one piece of construction paper within the following guidelines:
    - Each group has 30 minutes to complete construction and testing of their tower.
    - All towers must be able to stand freely for 10 seconds in order to qualify.
    - Prototype towers are to be constructed from the white paper only.
    - After testing the prototype tower and its ability to stand for 10 seconds, groups may make improvements to the design.
    - Construct the final tower from the construction paper.
    - Groups will get no more tape or paper during the challenge.
    - The final tower must be able to stand alone for at least 10 seconds with no type of attachment to the table.
    - The tallest tower free-standing for the longest period of time wins the challenge.

# Ten Second Tower
# Student Sheet

## Challenge Guidelines:

Construct the tallest free-standing tower possible from one piece of construction paper within the following guidelines:

- Each group has 30 minutes to complete construction and testing of their tower.
- All towers must be able to freely stand for 10 seconds in order to qualify.
- Prototype towers are to be constructed from the white paper only.

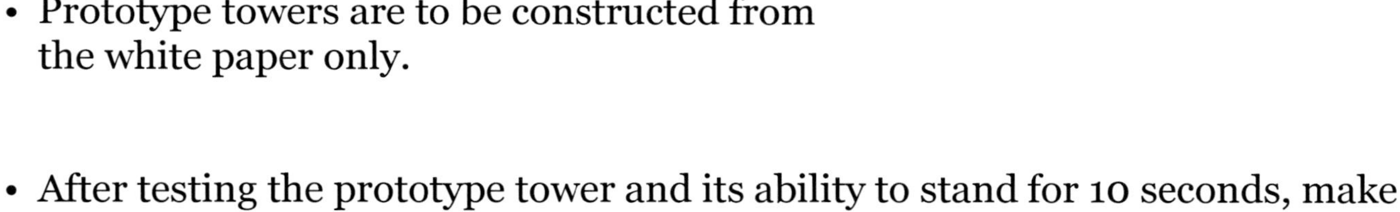

- After testing the prototype tower and its ability to stand for 10 seconds, make improvements to the design.
- Construct the final tower from the construction paper.
- Groups will get no more tape or paper during the challenge.
- The final tower must be able to stand alone for at least 10 seconds with no type of attachment to the table.
- The tallest tower free-standing for the longest period of time wins the challenge.

# Engineering Design Checklist
# Student Sheet

______ The structure meets the stated challenge requirements.

______ The project design team followed the challenge rules.

______ The project team exhibited effective, cooperative group work with every member participating.

______ The project prototype was carefully designed and built (if required).

______ The project team showed perseverance and a willingness to try again if necessary.

______ The project team was inventive in redesigning the original prototype.

______ The final structure shows creativity and originality.

______ The project meets the stated goals of the team.

# The Two Feet Feat

## Extending the Tower Challenge

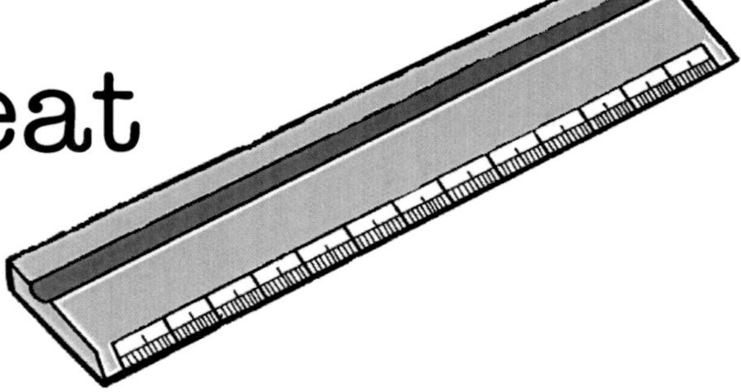

Time Required: 30 minutes plus discussion

### How this Learning Experience Meets the National Science Education Standards:

As a result of activities in grades 5-8, all students should develop:

### Content Standard E: Science and Technology

Abilities of Technological Design:

- Design a solution or product.
- Implement a proposed design.
- Evaluate completed technological designs or products.
- Communicate the process of technological design.

Understanding Science and Technology:

- Scientific inquiry and technological design have similarities and differences.
- Technological designs have constraints.
- Technological solutions have intended benefits and unintended consequences.

### Content Standard G: History and Nature of Science

Science as a Human Endeavor

- Women and men of various social and ethnic backgrounds engage in activities of science, engineering, and related fields; some scientists work in teams, and some work alone, but all communicate extensively with others.
- Science requires different abilities, depending on such factors as the field and study and the type of inquiry.

Nature of Science

- Scientists formulate and test their explanations of nature using observation, experiments, and theoretical and mathematical models.
- It is part of the scientific inquiry to evaluate the results of scientific investigations, experiments, observations, theoretical models, and the explanations proposed by other scientists.

## Teacher Notes:

The following concepts can be developed fully through an exploration in engineering design:

- Forces
- Torque
- Center of mass
- Stability
- Tension
- Compression
- Beam strength
- Column strength

## Safety Notes:

Participants should be reminded of the safe handling of scissors and pins; in addition, the work area should be clear of obstacles. Movement throughout the work area should be careful.

## Getting Started:

1. Collect the materials needed for the challenge.
2. Determine the size of the groups engaging in the experience.
3. Identify a work area that can accommodate the construction and testing of the straw towers.
4. Copy student challenge sheets.
5. Copy Engineering Design checklists, if needed.

## Materials Needed Per Group of Students:

- 50 plastic straws
- 50 straight pins or paper clips
- Scissors
- Weights, washers, or film canisters filled with sand
- Large paper clips to attach weights
- Meter stick
- Balance or scale (if needed to determine weight of washers/film canisters)
- Student challenge sheets
- Engineering design checklists (if appropriate)

## Procedure:

1. Assemble students into cooperative groups.
2. Challenge each group to construct the strongest two foot tower from straws and pins within the following guidelines:
   - Each group has 30 minutes to complete construction and testing of their tower.
   - All towers must be two feet only.
   - Towers must be free-standing and not anchored to the table in any way or leaning against any other structure.
   - Each tower's strength will be tested by suspending a series of weights from the center of the structure.
   - The tower must be able to stand freely with the weights attached for 30 seconds.
   - Groups will get no more straws or pins/paper clips during the challenge.
   - The tower that can hold the most weight will wins the challenge.

# The Two Feet Feat
# Student Sheet

## Challenge Guidelines:

Construct a strong two-foot tower from straws and pins within the following guidelines:

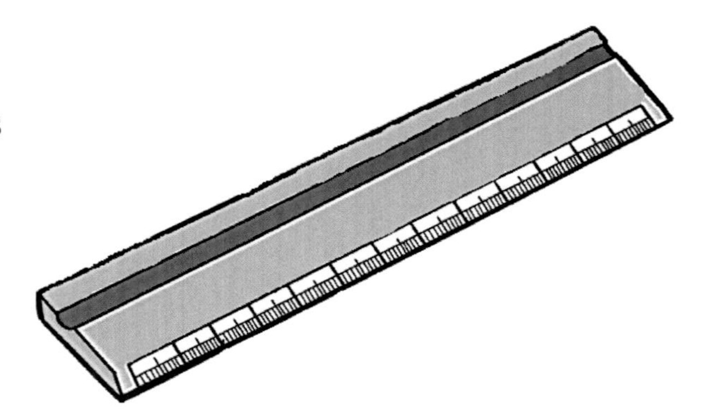

- Each group has 30 minutes to complete construction and testing of their tower.
- All towers must be two feet only.
- Towers must be free-standing and not anchored to the table in any way or leaning against any other structure.
- Each tower's strength will be tested by suspending a series of weights from the center of the structure.
- The tower must be able to stand freely with the weights attached for 30 seconds.
- Groups will get no more straws or pins/paper clips during the challenge.
- The tower that can hold the most weight will wins the challenge.

# Engineering Design Checklist
# Student Sheet

______ The structure meets the stated challenge requirements.

______ The project design team followed the challenge rules.

______ The project team exhibited effective, cooperative group work with every member participating.

______ The project prototype was carefully designed and built (if required).

______ The project team showed perseverance and a willingness to try again if necessary.

______ The project team was inventive in redesigning the original prototype.

______ The final structure shows creativity and originality.

______ The project meets the stated goals of the team.

# BOAT BONANZA!

## Exploring Boats and More

Time Required: 90-120 minutes; an additional 60-90 minutes is needed for the large-scale construction challenge

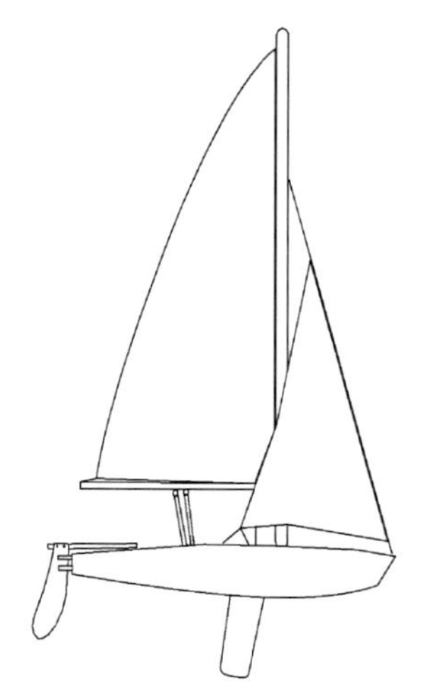

## How this Learning Experience Meets the National Science Education Standards:

As a result of activities in grades 5-8, all students should develop:

### Content Standard A: Science as Inquiry

Abilities Necessary to Do Scientific Inquiry

- Identify questions that can be answered through scientific investigations.
- Design and conduct a scientific investigation.
- Use appropriate tools and techniques to gather, analyze and interpret data.
- Develop descriptions, explanations, predictions and models using evidence.
- Think critically and logically to make the relationships between evidence and explanations.
- Recognize and analyze alternative explanations and predictions.
- Communicate scientific procedures and explanations.
- Use of mathematics in all aspects of scientific inquiry.

Understanding Scientific Inquiry

- Different kinds of questions suggest different kinds of scientific investigations.
- Mathematics is important in all aspects of scientific inquiry.
- Technology used to gather data enhances accuracy and allows scientists to analyze and quantify results of investigations.
- Scientific explanations emphasize evidence, have logically consistent arguments, and use scientific principles, models, and theories.
- Scientific investigations sometimes result in new ideas and phenomena for study, generate new methods or procedures for an investigation, or develop new technologies to improve the collection of data.

### Content Standard B: Physical Science

Motions and Forces

- Unbalanced forces will cause changes in the speed or direction of an object's motion.

### Content Standard E: Science and Technology

Abilities of Technological Design:

- Design a solution or product.
- Implement a proposed design.
- Evaluate completed technological designs or products.
- Communicate the process of technological design.

### Understanding Science and Technology:

- Scientific inquiry and technological design have similarities and differences.
- Technological designs have constraints.
- Technological solutions have intended benefits and unintended consequences.

## Content Standard G: History and Nature of Science

### Science as a Human Endeavor

- Women and men of various social and ethnic backgrounds engage in activities of science, engineering, and related fields; some scientists work in teams, and some work alone, but all communicate extensively with others.
- Science requires different abilities, depending on such factors as the field and study and the type of inquiry.

### Nature of Science

- Scientists formulate and test their explanations of nature using observation, experiments, and theoretical and mathematical models.
- It is part of the scientific inquiry to evaluate the results of scientific investigations, experiments, observations, theoretical models, and the explanations proposed by other scientists.

### History of Science

- In looking at the history of many peoples, one finds that scientists and engineers of high achievement are considered to be among the most valued contributors to their culture.
- Tracing the history of science can show how difficult it was for scientific innovators to break through the accepted ideas of their time to reach the conclusions that we currently take for granted.

## Objective:

Students work cooperatively to construct boats as well as identify the factors that determine how many passengers the boat will hold before sinking. Students will discover the relationship between the volume of a boat and the number of passengers it can safely carry, and then be able to predict the number of passengers a boat can carry if given its volume.

## Teacher Notes:

It is critical to remember that the pennies used as "passengers" in the learning experience MUST be dated 1983 and after. DO NOT USE PENNIES THAT PRE-DATE 1983. The content of pennies changed in 1982, so using pennies after that date will keep the masses of the pennies as similar as possible.

If you choose to do the Extension Challenge, you may want to consider testing different containers to see which one will be the most effective tool when assessing the final product. Because students are allowed to use a full sheet of poster board, the final boat product may be too large for the bowl or bucket that you have chosen to use previously in the experience. It is helpful if you have made a boat from poster board yourself and understand the process involved. Should time permit and facilities are available, an excellent follow-up to the extension is to challenge students to construct a boat from cardboard and tape. The objective is to see which boat can hold the most weight while being paddled across a larger body of water (swimming pool, pond, etc.). A student sheet is provided for this additional challenge.

## Safety Notes:

Because water is being used, have adequate supplies to clean up spills. Caution students about moving carefully around the work tables and to be aware of wet areas on the floor. Exercise care when working with the scissors; have a first aid kit available in case of cuts.

## Getting Started:

1. Gather and organize the materials needed for the series of learning experiences; based on student needs and class objectives, determine whether students will engage in the entire series or select parts.
2. If using the film "Titanic" as an engaging experience, obtain a copy and prepare the appropriate clip that reflects the crisis that resulted when trying to load lifeboats as the ship went down; identify the necessary technology needed to present the film clip to the students.
3. Determine whether students will work in full cooperative groups or in pairs within the groups.
4. If concepts related to the experiences are to be developed, identify appropriate content to be addressed.
5. Copy student sheets as needed.
6. Gather graphing materials; if a class graph is to be constructed, identify the appropriate means for displaying the data collected.
7. Determine whether students will engage in the boat construction extension; gather materials as needed to complete the experience.
8. Work with the different components of the experience yourself in order to become familiar with the testing process as well as the possible scenarios that the students may encounter; consider performing demonstrations at different points in order to better facilitate student learning.

## Materials Needed Per Group (or Pair) of Students:

- 4 paper cups
- 1 small plastic bowl
- Permanent marker
- Paper towels/sponge
- Large bowl or bucket of water
- Student sheets
- 1 plastic cup
- Pencil
- Scissors
- Ruler
- Graduated cylinder
- Learning logs
- 50 pennies *(must be from 1983 or newer to ensure masses are the same)
- Graph paper or boards
- Materials to prepare a class graph (optional)
- Chalkboard or overhead projector with transparency

## Materials Needed Per Group (or Pair) of Students for the Extension:

- Scissors
- Meter stick
- Pencil
- 1 piece of poster board
- Roll of duct tape

## Additional Materials Needed for Extension:

- Washers
- Extra-large bucket of water

## Procedure:

1. If using the engaging experience, have students remain in a whole group setting; instruct students to have learning logs ready.
2. Instruct students to carefully watch the following film clip and to make notes as needed.
3. Show a short clip from "Titanic" that illustrates a problem with the loading of available lifeboats; repeat if needed.
4. Pose the following questions to the students and have them record answers in learning logs: What was the problem with the lifeboats on the Titanic? What is the purpose of lifeboats? How are they used? What are some other types of boats and their uses?
5. Facilitate a discussion based on student responses; inform students that they will now explore their own "cup boats" and "penny passengers."
6. Have students assemble in cooperative groups.
7. Provide students with a copy of "Boat Bonanza, Part 1;" review the instructions with the groups.
8. Instruct the materials managers to obtain the necessary supplies for the exploration; review safety as needed before engaging students in the experience as you monitor their work.
9. Allow the students time to complete the process at least twice within the exploration.
10. Facilitate a discussion that reflects the following:
    - Have each group share the maximum number of "penny passengers" that their boat would hold.
    - Challenge students to develop reasons why the results varied when all of the "cup boats" were the same size.

- On the board or overhead, list all of the factors (or variables) that the students believe could have affected the maximum number of "penny passengers;" possibilities include "cup boats" are not measured to exactly three cm or there was some other measurement error, conditions of the "bucket pond" when someone bumped the table or shook the water, the way the passengers were added (carefully placed or dropped), the arrangement and position of the passengers in the boat.
- Engage the whole group in developing a standard method of loading passengers into the boats that will be adhered to for the remaining of the experience.

11. Provide students with copies of "Boat Bonanza, Part 2;" allow students the time needed to determine the capacity of their boats.
12. Following the completion of Part 2, compare the capacity of the "cup boats" to the original number of "penny passengers" that was initially reported by each group; guide students to the discovery that the slightly larger boats held more passengers.
13. Provide groups with a copy of "Boat Bonanza, Part 3."
14. Allow students time to construct new "cup boats" before predicting how many "penny passengers" it will hold, and then test their predictions.
15. When the groups have finished with their own boats, have them trade their two new boats with another group, and repeat the process with the partner boats.
16. Challenge students to compare with their trading partners in order to determine if both groups got the same maximum number of penny passengers.
17. Prepare a class graph on which the groups can plot their data. Put the volume of the cup boat in milliliters on the x-axis (independent variable) and the maximum number of penny passengers held on the y-axis (dependent variable).
18. When the graph is complete, facilitate an active discussion that includes the following:
    - Which "cup boat" held the most "penny passengers"?
    - What characteristic allowed it to hold the most?
    - What relationship was determined to exist between the volume of the cup boat and the number of penny passengers it could carry?
    - Using the graph, could you now predict the number of "penny passenger" that different size "cup boats" could hold?
    - Give the students three "cup boat" volumes and have them use the graph to predict the number of "penny passengers" each could carry.
19. If appropriate, engage students in the extension activity as follows:

    using the materials provided, construct a boat that will float in the bucket pond and hold the most washers before sinking.
20. Allow students ample time to complete construction; judge all boats according to the criteria listed before announcing a winner.
21. Facilitate a discussion that highlights the boat construction and factors that contributed to its success.

# Boat Bonanza
# Part One Instructions
## Student Sheet

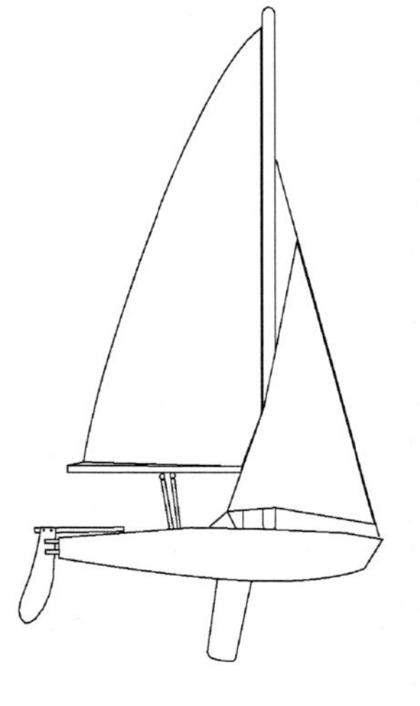

1. Using your ruler and pencil, mark a place three cm from the bottom of your first cup boat; cut it off at this height.

2. Test to be sure that your "cup boat" floats in its "bucket pond."

3. Add "penny passengers" to find out how many your cup boat can hold before sinking. (You must actually SINK your boat to determine the correct number.) Repeat the process to determine the most effective method of loading penny passengers.

4. Record the maximum number of penny passengers your cup boat held before sinking. _________

5. Describe the method of loading penny passengers that worked best for your group.

6. Be prepared to share your method of loading and your maximum number of penny passengers held with the class.

# Boat Bonanza
# Part Two Instructions
# Student Sheet

1. Place your cup boat into the empty small plastic bowl.

2. Dip water from your "bucket pond" with the plastic cup and fill your cup boat to the rim with water. If any overflows into the bowl, soak it up with paper towels or a sponge.

3. Tip your full cup boat so that all of the water it held pours into the plastic bowl.

4. Pinch the edge of your bowl to make a "spout," and very carefully pour the water into your graduated cylinder. Record the volume (capacity) of your boat in milliliters.

   __________

5. Be prepared to share the capacity of your boat with the class.

# Boat Bonanza
# Part Three Instructions
# Student Sheet

1. Make two new boats, one smaller and one larger than your original. Do not make a boat larger than four cm tall.

2. Record your new boat data

| New Boat | Volume (mL) | Predicted Penny Passengers | Actual Penny Passengers |
|---|---|---|---|
| Small | | | |
| Large | | | |

3. Trade boats with another group and repeat the process.

| New Boat | Volume (mL) | Predicted Penny Passengers | Actual Penny Passengers |
|---|---|---|---|
| Small | | | |
| Large | | | |

4. Compare "maximums" with your trade partners.

5. Put your own group data on the class graph.

# Group Extension
# A Construction Challenge

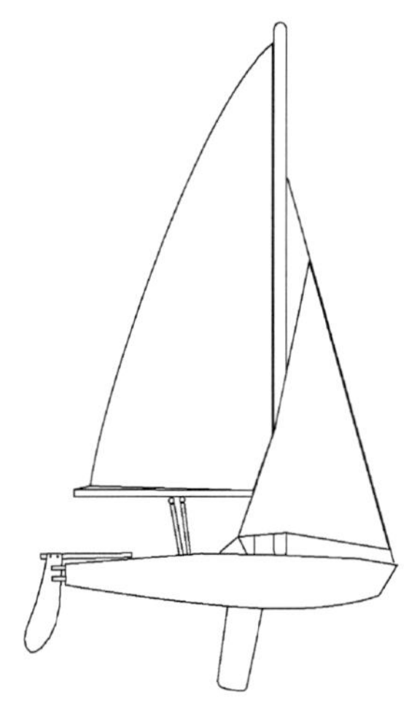

## Guidelines:

Using the following list of materials, your group is challenged to build the best boat possible:

- Scissors
- 1 piece of poster board
- Meter stick
- Roll of duct tape
- Pencil

## Judging Criteria:

The best boat will float in the "bucket pond" while holding the most washers before sinking. The washer that sinks the boat will be counted in the total.

Your time limit is 30 minutes.

Good luck!!

# Boat Bonanza
# A Large Scale Construction Challenge

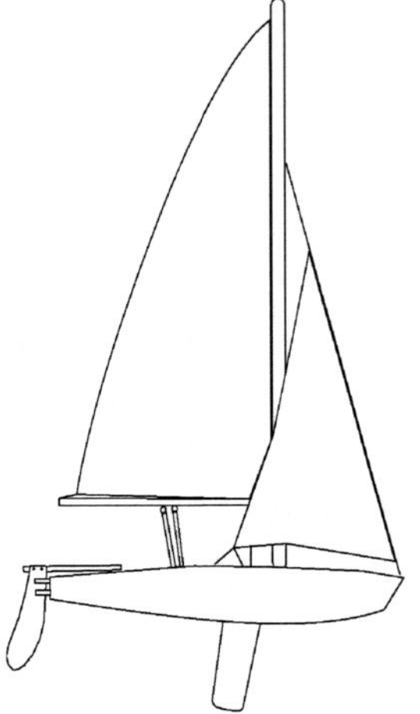

## Challenge Guidelines:

Construct a cardboard boat within the following guidelines:

- Each group has 60-90 minutes to complete construction and testing of their boat.
- All chairs must be made from cardboard and duct tape only—no glue, staples, etc.
- The structure must resemble a boat.
- Boats must be able to free float and not attached to anything for support.
- Each group must be able to pick up its boat as one piece.
- The chair must support the weight of at least one of its group members while being paddled across water.
- The chair that can hold the most weight while being paddled across water will win the challenge.
- Should there be a tie, the boat that sinks last will be declared the winner.

# Civil Engineering Word Search Puzzle

O J F C P S J L D G P L B C J A Q P T T P B T M R

V P C P H X T E B E B A V H M T G V K S S W E C D

P S P T T P J Q L Y O T D A O A H Z D A R T U X C

W C J O C K F Q N W D N T E O S O T E T E P C A M

M H Y G R T P B I N N E F G T F G C Z X T Z E E W

P O X T L T Z Z G N R M E H E E S J O Z S U K J B

W L O P X Y U B S I R N Y C T A C M S P A X A A R

F A Y X G I B N A T C O I R K E J T E U O R U P A

Y R T M P V S L I W D R T A U X C W I M C F Q Z N

C S M J T R S Q K T T I A E L D Q H T V R C H C C

C H D H J L P Y E T Y V T S M L O T N G E P T O H

S I G R X L X A R V Y N O E Z C I N I O L S R U E

P P W Q V S P Z U W J E Z R J V N H V Q L B A P S

C S Q P C B S R E D R A O B E T A K S E O O E H R

K N O I T A T R O P S N A R T K Z L Y J R L G Y Z

H G V O N Z G E O T E C H N I C A L P X T W B Y P

1. They may design the framing and structure of ______________

2. As ______________ specialist. They may make traffic systems telling people when to stop or go.

3. Civil engineers may design portable ramps and ledges for ________________ learning to do tricks at home.

4. _____________ engineers are also know as pollution doctors.

5. ________________ engineers may help maintain the Hoover dam or construct seawalls.

6. ASCE supports students and encourages growth by offering numerous _________________.

7. AKA Ground ___________. They may learn about the earth's crust and make sure anything going into the ground is safe.

8. Some civil engineers opt for a career in _________.

9. They may understand steel, concrete, wood, and all sorts of other ______________.

10. A civil engineer who goes to law school could be an _________ or hurricane insurance attorney

11. As our current understanding of __________ increases, demand for the diverse talents of civil engineers will increase too.

12. Civil engineering is one of the oldest and largest _________ of engineering.

13. There is no limit to the versatility and ____________ of the civil engineering profession.

14. American Society of Civil Engineers

15. Approximately ____________ percent of civil engineers work in construction, transportation, manufacturing and utilities.

# Civil Engineering Word Search Puzzle
## Teacher Sheet

```
O G Z O M N D N N A H L D E Y P O C Q G L A U L P
K P U U K Y M G H Y K A W N M H S M F J S P F G D
G S P C J F C L V P Q T D A O K B V J T R U T V Z
W C X O T S K G T C Z N T E K O B H E L E S U Q P
O H Y P R Q P X A E I E A P T S L C Z F T N E K X
W O I T T T W D U N R M Y H C E S W N B S N K O B
X L I L X X U R P I O N N C T A C I Q B A F A U R
Q A Z A Z I L N A U H O Z R X E W T O R O H U G A
E R J I C J S L I H M R N A S I C X I M C S Q P N
I S V O U C S Z N T X I B E G Q Q H D V R F H C C
F H B P G B F E Z P Y V T S E M J K N X E O T J H
F I M U D Y S N C W V N R E T J W Q W O L S R M E
P P R N F L K T O R Y E K R S L R V C B L F A C S
I S S C V I S R E D R A O B E T A K S Z O O E K M
Y N O I T A T R O P S N A R T S K C I I R H G B V
K D E Y K E G E O T E C H N I C A L E F H J R Y F
```

1. They may design the framing and structure of _______________ [rollercoasters]

2. As _______________ specialist. They may make traffic systems telling people when to stop or go. [transportation]

3. Civil engineers may design portable ramps and ledges for _________________ learning to do tricks at home. [skateboarders]

4. ______________ engineers are also know as pollution doctors. [environmental]

5. _________________ engineers may help maintain the Hoover dam or construct seawalls. [geotechnical]

6. ASCE supports students and encourages growth by offering numerous __________________. [scholarships]

7. AKA Ground ____________. They may learn about the earth's crust and make sure anything going into the ground is safe. [detectives]

8. Some civil engineers opt for a career in _________. [research]

9. They may understand steel, concrete, wood, and all sorts of other _______________. [materials]

10. A civil engineer who goes to law school could be an _________ or hurricane insurance attorney [earthquake]

11. As our current understanding of __________ increases, demand for the diverse talents of civil engineers will increase too. [technology]

12. Civil engineering is one of the oldest and largest _________ of engineering. [branches]

13. There is no limit to the versatility and ____________ of the civil engineering profession. [opportunity]

14. American Society of Civil Engineers [ASCE]

# Civil Engineering Crossword Puzzle

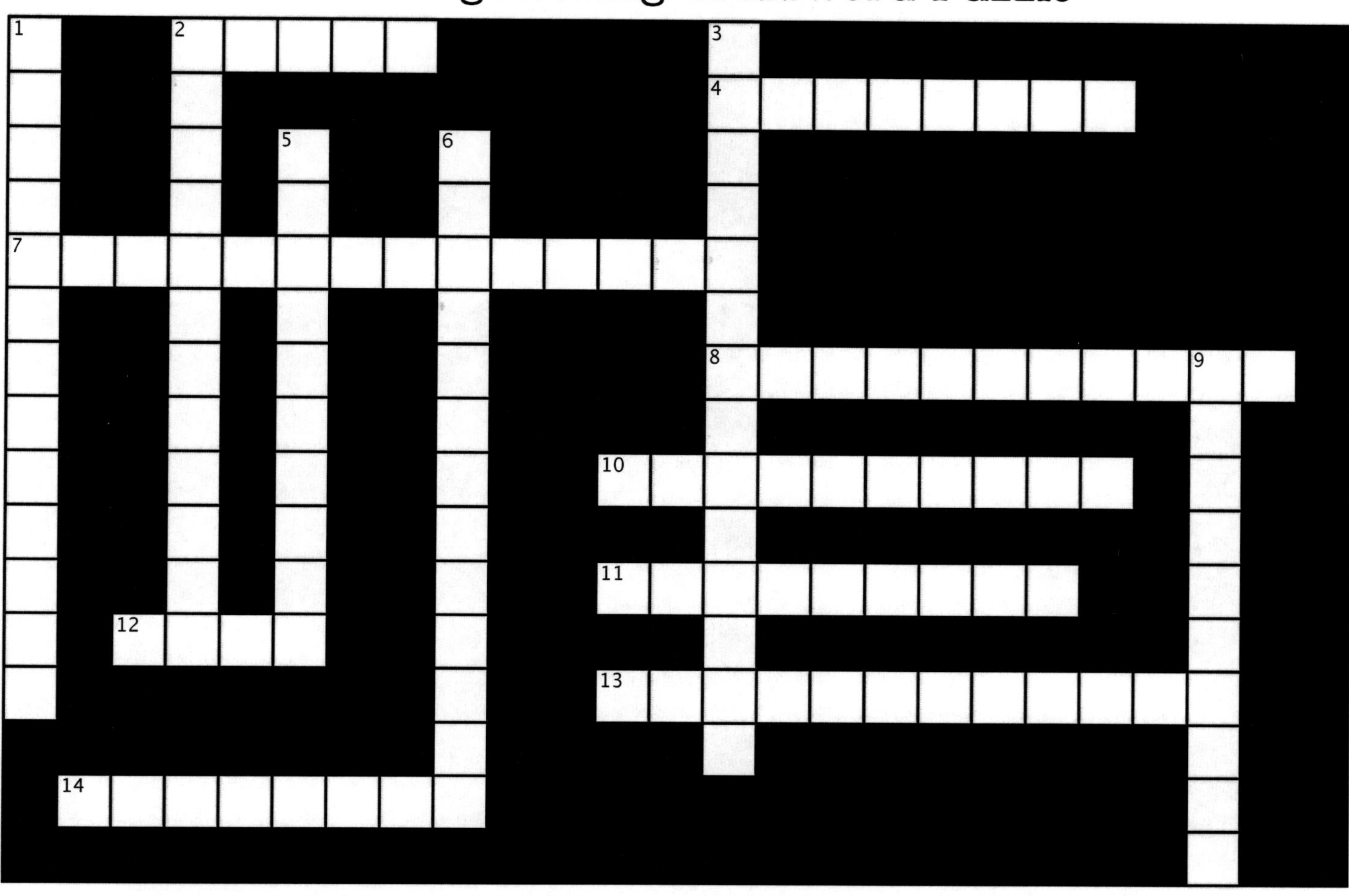


## Across

2. Approximately ____________ percent of civil engineers work in construction, transportation, manufacturing and utilities.
4. Some civil engineers opt for a career in _________.
7. They may design the framing and structure of ______________
8. There is no limit to the versatility and ____________ of the civil engineering profession.
10. AKA Ground ___________. They may learn about the earth's crust and make sure anything going into the ground is safe.
11. They may understand steel, concrete, wood, and all sorts of other _______________.
12. American Society of Civil Engineers
13. _________________ engineers may help maintain the Hoover dam or construct seawalls.
14. Civil engineering is one of the oldest and largest _________ of engineering.

## Down

1. _____________ engineers are also know as pollution doctors.
2. ASCE supports students and encourages growth by offering numerous __________________.
3. As _______________ specialist. They may make traffic systems telling people when to stop or go.
5. A civil engineer who goes to law school could be an _________ or hurricane insurance attorney
6. Civil engineers may design portable ramps and ledges for _________________ learning to do tricks at home.
9. As our current understanding of __________ increases, demand for the diverse talents of civil engineers will increase too.

# Civil Engineering Crossword Puzzle
## Teacher Sheet

| 1 E | | | 2 S | I | X | T | Y | | | | | | 3 T | | | | | | | | | | | |
|---|---|---|---|---|---|---|---|---|---|---|---|---|---|---|---|---|---|---|---|---|---|---|---|---|
| N | | | C | | | | | | | | | | 4 R | E | S | E | A | R | C | H | | | | |
| V | | | H | | 5 E | | | 6 S | | | | | A | | | | | | | | | | | |
| I | | | O | | A | | | K | | | | | N | | | | | | | | | | | |
| 7 R | O | L | L | E | R | C | O | A | S | T | E | R | S | | | | | | | | | | | |
| O | | | A | | T | | | T | | | | | P | | | | | | | | | | | |
| N | | | R | | H | | | E | | | | | 8 O | P | P | O | R | T | U | N | I | 9 T | Y | |
| M | | | S | | Q | | | B | | | | | R | | | | | | | | | E | | |
| E | | | H | | U | | | O | | | 10 D | E | T | E | C | T | I | V | E | S | | C | | |
| N | | | I | | A | | | A | | | | | A | | | | | | | | | H | | |
| T | | | P | | K | | | R | | | 11 M | A | T | E | R | I | A | L | S | | | N | | |
| A | | 12 A | S | C | E | | | D | | | | | I | | | | | | | | | O | | |
| L | | | | | | | | E | | | 13 G | E | O | T | E | C | H | N | I | C | A | L | | |
| | | | | | | | | R | | | | | N | | | | | | | | | O | | |
| | 14 B | R | A | N | C | H | E | S | | | | | | | | | | | | | | G | | |
| | | | | | | | | | | | | | | | | | | | | | | Y | | |

### Across

2. Approximately _____________ percent of civil engineers work in construction, transportation, manufacturing and utilities. [sixty]
4. Some civil engineers opt for a career in _________. [research]
7. They may design the framing and structure of _______________ [rollercoasters]
8. There is no limit to the versatility and _____________ of the civil engineering profession. [opportunity]
10. AKA Ground ____________. They may learn about the earth's crust and make sure anything going into the ground is safe. [detectives]
11. They may understand steel, concrete, wood, and all sorts of other _______________. [materials]
12. American Society of Civil Engineers [ASCE]
13. _________________ engineers may help maintain the Hoover dam or construct seawalls. [geotechnical]
14. Civil engineering is one of the oldest and largest _________ of engineering. [branches]

### Down

1. ______________ engineers are also know as pollution doctors. [environmental]
2. ASCE supports students and encourages growth by offering numerous __________________. [scholarships]
3. As _______________ specialist. They may make traffic systems telling people when to stop or go. [transportation]
5. A civil engineer who goes to law school could be an __________ or hurricane insurance attorney [earthquake]
6. Civil engineers may design portable ramps and ledges for __________________ learning to do tricks at home. [skateboarders]
9. As our current understanding of ___________ increases, demand for the diverse talents of civil engineers will increase too. [technology]

# Resources - Engineering Career Websites

## General Engineering

**Engineering Education Service Center -** www.engineeringedu.com
**Teach Engineering -** www.teachengineering.org
**Engineering Go For It** - www.egfi-K12.org

## Ceramic Engineering

**A Fiber Optics Experiment -** engineering.alfred.edu/fun_activities/flashlightfiberoptics.html
**A Musical Glass Experiment -** engineering.alfred.edu/fun_activities/musicalglass.html
**The Materials Science & Engineering Career Resource Center -** www.crc4mse.org

## Metallurgical Engineering

**Tug-Push-Twist-O'War -** Materials Lab Hands-On Activity Handout
www.pbs.org/wgbh/buildingbig/educator/material_war.html
**THE WILD WEST: All That Glitters – Teaching Guide**
www.pbs.org/safarchive/4_class/45_pguides/pguide_601/4561_glit.html

## Environmental Engineering

**National Geographic Online - A** content-rich site full of online articles, maps, geography quizzes, and more. You can chat with Society photographers, writers, and artists. You can even exchange ideas with the leading scientific minds of our time. Site also provides many lesson plans. www.nationalgeographic.com
**U.S. Environmental Protection Agency -** www.epa.gov/epahome/Programs.html
**Volcano World** is perfect for students at any level. It has timely updates about volcanic activity worldwide, historical eruption reports, information about how volcanoes work and guidance regarding becoming a volcanologist. volcano.und.nodak.edu
**WhaleNet** focuses on whales and marine research. It is dedicated to interdisciplinary education. Their goal is to foster excitement about learning and the environment. whale.wheelock.edu

## Systems Engineering

**It's So Simple** – This lesson may be used as an introduction to Simple Machines. www.iit.edu/~smilemp0298.htm
**Modeling the Nervous System** - Sometimes the best way to learn about something is to hold it in your hand. What better way to learn about the different parts of the nervous system than to make them yourself. faculty.washington.edu/chudler/chmodel.html
**Study the effects of friction in a simple mechanical system.** www.iit.edu/~smile/ph8623.html

## Electrical Engineering

**PEERS – PreCollege Educators/Engineers Resource Site**
This section provides links to a variety of national programs and projects that either provide support to teachers, or encourage students in the study of science, mathematics, engineering, and/or technology.
www.ieee.org/organizations/eab/precollege/peers/resources/prog.htm

# About the Authors

Celeste Baine is a biomedical engineer, director of the Engineering Education Service Center and the award winning author of over twenty books and booklets on engineering careers and education. She won the Norm Augustine Award from the National Academy of Engineering (The Norm Augustine award is given to an engineer who has demonstrated the capacity for communicating the excitement and wonder of engineering). She also won the American Society for Engineering Education's Engineering Dean Council's Award for the Promotion of Engineering Education and Careers and is listed on the National Engineers Week website as one of 50 engineers you should meet. The National Academy of Engineering has included Celeste in their Gallery of Women Engineers and she has been named one of the Nifty-Fifty individuals who have made a major impact on the field of engineering by the USA Science and Engineering Festival.

Cathi Cox left a 17 year career at Choudrant High School in 1998 to become the Site Coordinator for Louisiana Tech University's Project LIFE. Since then, her responsibilities have expanded into a position as Program Coordinator for Tech's CATALyST (Center for Applied Teaching and Learning to Yield Scientific Thinking). Currently serving in that capacity as well as the Site Coordinator for C3 (Chemical Concepts and Connections for Teacher Leaders), Cathi has worked passionately to become a highly successful facilitator of presentations that focus on reform methodologies and strategies, alternative assessment, school to work, and standards based education. Her enthusiastic approach has been shared through summer projects, methods courses, workshops, seminars, and online classes throughout the state and southern region.